Alina Filatova

Nuclear-cytoplasmic translocation of the transporter regulator RS1

Alina Filatova

Nuclear-cytoplasmic translocation of the transporter regulator RS1

Mechanism and control

Südwestdeutscher Verlag für Hochschulschriften

Impressum / Imprint
Bibliografische Information der Deutschen Nationalbibliothek: Die Deutsche Nationalbibliothek verzeichnet diese Publikation in der Deutschen Nationalbibliografie; detaillierte bibliografische Daten sind im Internet über http://dnb.d-nb.de abrufbar.
Alle in diesem Buch genannten Marken und Produktnamen unterliegen warenzeichen-, marken- oder patentrechtlichem Schutz bzw. sind Warenzeichen oder eingetragene Warenzeichen der jeweiligen Inhaber. Die Wiedergabe von Marken, Produktnamen, Gebrauchsnamen, Handelsnamen, Warenbezeichnungen u.s.w. in diesem Werk berechtigt auch ohne besondere Kennzeichnung nicht zu der Annahme, dass solche Namen im Sinne der Warenzeichen- und Markenschutzgesetzgebung als frei zu betrachten wären und daher von jedermann benutzt werden dürften.

Bibliographic information published by the Deutsche Nationalbibliothek: The Deutsche Nationalbibliothek lists this publication in the Deutsche Nationalbibliografie; detailed bibliographic data are available in the Internet at http://dnb.d-nb.de.
Any brand names and product names mentioned in this book are subject to trademark, brand or patent protection and are trademarks or registered trademarks of their respective holders. The use of brand names, product names, common names, trade names, product descriptions etc. even without a particular marking in this work is in no way to be construed to mean that such names may be regarded as unrestricted in respect of trademark and brand protection legislation and could thus be used by anyone.

Verlag / Publisher:
Südwestdeutscher Verlag für Hochschulschriften
ist ein Imprint der / is a trademark of
OmniScriptum GmbH & Co. KG
Heinrich-Böcking-Str. 6-8, 66121 Saarbrücken, Deutschland / Germany
Email: info@svh-verlag.de

Herstellung: siehe letzte Seite /
Printed at: see last page
ISBN: 978-3-8381-1161-2

Zugl. / Approved by: Wuerzburg, University of Wuerzburg, Diss., 2009

Copyright © 2009 OmniScriptum GmbH & Co. KG
Alle Rechte vorbehalten. / All rights reserved. Saarbrücken 2009

Table of contents

1. INTRODUCTION 5

1.1. The RS1 protein 5
1.2. Transport of proteins in and out of the nucleus 8
1.2.1. Nuclear protein import pathways 9
1.2.2. Nuclear export pathways 12
1.2.3. Regulation of nuclear transport 13
1.2.3.1. Regulation of the karyopherin-cargo interaction 13
1.2.3.2. Regulation of transport by the NPC 15

2. THE AIM OF THIS STUDY 17

3. MATERIALS AND METHODS 18

3.1. Materials 18
3.1.1. Chemicals 18
3.1.2. Antibodies 18
3.1.3. Affinity matrices 19
3.1.4. DNA and Protein Markers 19
3.1.5. Enzymes 19
3.1.6. Inhibitors and activators 19
3.1.7. Reaction kits 20
3.1.8. Peptides 20
3.1.9. Synthetic Oligonucleotides 20
3.1.10. Plasmids and constructs 21
3.1.11. Bacteria 24
3.1.12. Cell lines 24
3.1.13. Buffers and solutions 24
3.1.14. Software 24
3.2. Methods 25
3.2.1. Molecular biology 25
3.2.1.1. Mutagenesis 25
3.2.1.2. Generation of the GFP-TEV-S-Tag vector 25
3.2.1.3. Generation of the GFP-TEV-S-Tag-CK2-NS-PKC-PKC vector 25
3.2.1.4. Annealing of oligonucleotides 26

3.2.1.5.	Polymerase chain reaction (PCR)	26
3.2.1.6.	DNA isolation by phenol/chloroform extraction	26
3.2.1.7.	Digestion of DNA with restriction endonucleases	27
3.2.1.8.	Analytical agarose gel electrophoresis of DNA	27
3.2.1.9.	Preparative agarose gel electrophoresis	27
3.2.1.10.	Ligation	28
3.2.1.11.	Analytical agarose gel electrophoresis of RNA	28
3.2.1.12.	Desalting of DNA samples	28
3.2.1.13.	Transformation of bacteria and clone selection	29
3.2.1.14.	Isolation of plasmid DNA from E.coli	29
3.2.1.15.	Determination of DNA and RNA concentration by spectrophotometry	30
3.2.2. Protein analysis methods		30
3.2.2.1.	Preparation of the whole-cell extracts	30
3.2.2.2.	Purification of GFP fusion proteins	30
3.2.2.3.	Immunoprecipitation of GFP fusion proteins and associated proteins	31
3.2.2.4.	Determination of protein concentration	32
3.2.2.5.	SDS-polyacrylamide gel electrophoresis	32
3.2.2.6.	Western blot and immunodetection	33
3.2.2.7.	Staining of protein polyacrylamide gels with Coomassie blue	34
3.2.2.8.	Silver staining of protein polyacrylamide gels	34
3.2.2.9.	Gel drying	35
3.2.3. Generation and testing of phosphospecific antibodies		35
3.2.3.1.	Immunization of rabbits	36
3.2.3.2.	Identification of the antibody titer. Enzyme-linked Immunosorbent Assay (ELISA)	37
3.2.3.3.	Affinity purification of antibodies	37
3.2.4. Cell Culture		38
3.2.4.1.	Cultivation of mammalian cells	38
3.2.4.2.	Passage	38
3.2.4.3.	Cryoculture	38
3.2.4.4.	Transient transfection of mammalian cells	39
3.2.4.5.	Generation of stable cell lines	39
3.2.4.6.	Inhibitor treatment of cells	39
3.2.5. Analysis of gene expression in mouse embryonic fibroblasts (MEFs)		40
3.2.5.1.	Isolation and cultivation of MEFs	40
3.2.5.2.	Cryoculture	41

3.2.5.3. Synchronization of MEFs 41
3.2.5.4. Isolation of Total RNA 41
3.2.5.5. Gene expression microarray analysis 41
3.2.6. Fluorescence analysis and measurements of nuclear localization 42
3.2.7. Calculation and Statistics 42

4. RESULTS 44

4.1. Analysis of nuclear location of pRS1 and its fragments: experimental design 44
4.2. Dynamic redistribution of pRS1 during the cell cycle 44
4.3. Identification and characterization of nuclear export signal in pRS1 46
4.4. hRS1 interacts with nuclear import receptor importin β1 49
4.5. Identification of a minimal sequence steering confluence dependent location of pRS1 .. 50
4.6. Investigation of the role of phosphorylation of serine 370 52
4.7. Studies on the phosphorylation state of serine 370 of pRS1 in subconfluent and confluent LLC-PK$_1$ cells using mass spectrometry 55
4.8. Investigation of the role of calmodulin in the regulation of nuclear location of pRS1 58

5. DISCUSSION 62

APPENDIX I. GENE EXPRESSION PROFILING IN RS1 DEFICIENT MOUSE EMBRYONIC FIBROBLASTS. 68

APPENDIX II. STUDIES ON THE UBIQUITINATION OF RS1. 71

APPENDIX III. STUDIES ON THE DEGRADATION OF RS1 IN SUBCONFLUENT AND CONFLUENT LLC-PK1 CELLS. INVESTIGATION OF THE ROLES OF THE PROTEASOME AND CALPAIN. 74

APPENDIX IV. GENERATION OF ANTIBODIES WHICH RECOGNIZE PHOSPHORYLATED SERINE 370 84

6. SUMMARY 88

8. ABBREVIATIONS 89

9. REFERENCES 90

ACKNOWLEDGEMENTS 105

1. Introduction

1.1. The RS1 protein

After cloning and functional characterization of a variety of solute transporters in the plasma membrane, the factors regulating their expression and functional activities have become an important topic in transporter research. Transporters can be regulated at the levels of transcription, mRNA stability and translation as well as at posttranslational level.

RSC1A1 is an intronless single copy gene that is specific for mammals and encodes 67-to 68-kDa RS1 proteins in human (Lambotte *et al.*, 1996), pig (Veyhl *et al.*, 1993), rabbit (Reinhardt *et al.*, 1999), and mouse (Osswald *et al.*, 2005). A first RS1 ortholog from pig was isolated by screening of an expression library from porcine kidney with a monoclonal IgM antibody which stimulated high affinity phlorizin binding to renal border membranes but did not react with Na^+-D-glucose cotransporter SGLT1 (Veyhl *et al.*, 1993). After co-expression of RS1 with SGLT1 in *Xenopus laevis* oocytes, the expressed uptake of methyl-α-D-glucopyranoside (AMG) and the apparent substrate dependence of AMG uptake after 1 hour incubation with substrate were altered. Moreover, immunohystochemical studies showed that RS1 was associated with the plasma membrane (Veyhl *et al.*, 1993;Lambotte *et al.*, 1996). On the basis of these data, RS1 was proposed to be a regulatory subunit of SGLT1 (Koepsell and Spangenberg, 1994). However, later this hypothesis appeared to be improbable since RS1 was shown to localize intracellularly in *Xenopus laevis* oocytes (Valentin *et al.*, 2000) and to regulate different membrane transporters (Veyhl *et al.*, 1993;Lambotte *et al.*, 1996;Veyhl *et al.*, 2003;Veyhl *et al.*, 2006).

RS1 orthologs exhibit about 70% identity on the amino acid sequence level. Several functional domains of the RS1 protein have been identified: (i) an N-terminal domain responsible for the posttranscriptional down-regulation of SGLT1, containing three regulatory peptides, two consensus sequences for potential binding to the 14-3-3 proteins, and two consensus PKC phosphorylation sites (Vernaleken *et al.*, 2007; Veyhl M., Vernaleken A., Koepsell H., unpublished data); (ii) a nuclear localization sequence (Leyerer, 2007), and (iii) a C-terminal ubiquitin associated domain (UBA) (Valentin *et al.*, 2000).

RS1 has a broad tissue distribution, including renal proximal tubules, small intestine, liver, and neurons (Veyhl *et al.*, 1993;Lambotte *et al.*, 1996;Poppe *et al.*, 1997;Reinhardt *et al.*, 1999;Valentin *et al.*, 2000). Low expression of RS1 was also detected in the lung and spleen but not in skeletal and heart muscle, colon or stomach (Veyhl *et al.*, 1993;Reinhardt *et al.*, 1999). In porcine kidney RS1 protein was found in brush-border membrane fraction (Valentin *et al.*, 2000). In mouse small intestine RS1 was localized in epithelial and subepithelial cells, within the nucleus and below the plasma membrane (Osswald et al., 2005). In renal porcine epithelial cell line LLC-PK$_1$, RS1 is located at the intracellular side of the plasma membrane, at the *trans*-Golgi netwotk (TGN), and within the nucleus (Kroiss et al., 2006). At that, nuclear location of RS1 is confluence-dependent. The RS1 protein is localized in the nuclei and cytoplasm of subconfluent LLC- PK$_1$ cells and in the cytoplasm of confluent LLC- PK$_1$ cells. In *Xenopus laevis* oocytes, most of the over-expressed RS1 protein is found in the cytosol; a small fraction of RS1 is also associated with the plasma membrane (Valentin et al., 2000).

Co-expression experiments in *Xenopus laevis* oocytes have shown that RS1 regulates the activities of SGLT1 and some other plasma membrane transporters including the SGLT1-homologous Na$^+$-myo-inositol cotransporter SMIT, the organic cation transporters OCT1 and OCT2, and the organic anion transporter OAT1 (Lambotte et al., 1996;Reinhardt et al., 1999;Veyhl et al., 2003). Apparently, RS1 can regulate transporters from different families. Although the selectivity of RS1 is not fully understood, it has been demonstrated that SGLT1 is a physiologically important target of RS1. RS1$^{-/-}$ mice developed obesity associated with increases in food intake, glucose transport and SGLT1 expression in the small intestine (Osswald et al., 2005). The effect of RS1 deficiency was tissue-specific, and downregulation of SGLT1 by RS1 in small intestine occurred through posttranscriptional mechanisms. These observations initiated a more detailed investigation of the role of the RS1 protein in the regulation of SGLT1.

The posttranscriptional regulation of SGLT1 by RS1 was studied in *Xenopus laevis* oocytes. Co-expression of hRS1 and hSGLT1 (Veyhl et al., 1993;Lambotte et al., 1996;Reinhardt et al., 1999;Veyhl et al., 2003) or injection of the purified RS1 protein into SGLT1 expressing oocytes (Veyhl et al., 2006) led to inhibition of SGLT1-mediated AMG uptake. The short-term posttranscriptional down-regulation of SGLT1 by RS1 occurred within 30 min and was due to blockage of the dynamin-dependent release of hSGLT1 containing vesicles from the TGN (Veyhl et al., 2006;Kroiss et al., 2006). This posttranscriptional down-regulation of SGLT1 by RS1 is increased upon activation of PKC and decreased at enhanced intracellular AMG concentration (Veyhl et al., 2006). Interestingly, the short-term inhibition of hOCT2-mediated tetraethylammonium uptake by hRS1 protein is decreased at high intracellular AMG concentration as well (Veyhl et al., 2006). The data suggests that inhibition of the transporters by RS1 is regulated by an intracellular glucose binding protein.

RS1 interacts with a recently identified 28-kDa ischemia/reperfusion inducible protein (IRIP), which is up-regulated in kidney after ischemia and reperfusion (Jiang et al., 2005). IRIP protein is expressed at relatively high levels in the testis, bronchial epithelia, thyroid, ovary, colon, kidney, and brain, and at the low levels in the spleen, muscle, heart, and small intestine. IRIP inhibits the expression of a variety of plasma membrane transporters including SGLT1, the organic cation transporters OCT2 and OCT3, the organic anion transporter OAT1, the Na^+-cotransporter for serotonin SERT, the dopamine transporter DAT, and the norepinephrine transporter NET. Interestingly, most of them have been shown to be regulated by RS1 as well (Lambotte et al., 1996;Reinhardt et al., 1999;Veyhl et al., 2003). Therefore, RS1 and IRIP can belong to a novel regulatory pathway that controls activities of the solute carriers of several families. This assumption is confirmed by experimental data. No additive or synergic interaction between effects of IRIP and RS1 on OCT2 was observed, and the effect of RS1 was abolished when the dominant negative mutant of IRIP was co-expressed (Jiang et al., 2005). Because RS1 and IRIP are expressed in various tissues, the regulatory pathway is supposed to be present in many cell types.

Attempts to identify a domain of hRS1 that is responsible for the posttranscriptional inhibition of SGLT1 were undertaken (Vernaleken et al., 2007). Two tripeptides derived from hRS1 sequence, GlnCysPro and GlnSerPro, were identified which act as high affinity posttranscriptional inhibitors of hSGLT1. Similar to the full-length hRS1 protein (Veyhl et al., 2006), GlnCysPro and GlnSerPro inhibit the release of hSGLT1 containing vesicles from the TGN. Moreover, the down-regulation of SGLT1 by the tripeptides is also modulated by different intracellular monosaccharides. Therefore, the mechanism of the tripeptide-mediated inhibition was proposed which involves binding of the tripeptides to a high affinity binding site of a protein, which contains a modulatory monosaccharide binding site, at the TGN (Vernaleken et al., 2007).

The similarities between the observed posttranscriptional effects of hRS1 and the tripeptides indicate that GlnCysPro and/or GlnSerPro form a part of (a) posttranscriptionally active domain(s) of hRS1. The domain responsible for the posttranscriptional down-regulation of SGLT1 has recently been identified (Veyhl, Vernaleken, Koepsell, unpublished data). It consists of two GlnSerPro motifs (aa 19-21, 91-93), another regulatory peptide, two consensus sequences for the binding of protein 14-3-3 and two consensus sequences for PKC-dependent phosphorylation. Since RS1 protein is rapidly degraded in HEK 293 cells, Caco-2 cells and LLC-PK_1 cells (Koepsell et al., unpublished data), and small hRS1 fragments including tripeptides demonstrate inhibitory activity, it is possible that the posttranscriptional down-regulation of hSGLT1 by hRS1 is partially mediated by hRS1 fragments. This question is still open and requires further investigation.

RS1 was suggested to participate in the transcriptional down-regulation of SGLT1 (Korn et al., 2001). When RS1 expression in porcine LLC-PK_1 cells was reduced via an antisense strategy, the expression of SGLT1 was up-regulated on the transcriptional and posttranscriptional levels; conversely, overexpression of RS1 caused a strong decrease in the expression of SGLT1. This inverse relationship between RS1 and SGLT1 gave rise to the hypothesis that RS1 suppresses transcription of SGLT1. Interestingly, SGLT1 expression is confluence-dependent in LLC-PK_1 cells. SGLT1 is virtually undetectable in subconfluent cells and is highly expressed in confluent cells. Thus, a hypothesis was raised that RS1 inhibits the expression of SGLT1 in subconfluent LLC-PK$_1$ cells and that the up-regulation of SGLT1 after confluence is caused by a relief of this inhibition.

Remarkably, the expression level and the distribution of RS1 are dependent on the state of confluence. Whereas subconfluent LLC-PK$_1$ cells contain large amounts of RS1 protein and exhibit pronounced nuclear location of RS1, in the confluent LLC-PK$_1$ cells the amount of RS1 is decreased and RS1 is located outside of the nucleus (Korn *et al.*, 2001;Kroiss *et al.*, 2006). The mechanisms underlying these changes might represent the regulatory mechanisms which govern RS1 function in LLC-PK$_1$ cells. For example, the observation that RS1 is located in the nucleus of subconfluent but not confluent LLC-PK$_1$ cells correlates with functional data showing that RS1 down-regulates the transcription of SGLT1 in subconfluent LLC-PK$_1$ cells (Korn *et al.*, 2001). The investigation of the regulation of dynamic RS1 localization and expression in relation to cell confluence can provide important new insights for the understanding of RS1 function which involves coordinated transcriptional and posttranscriptional regulation.

1.2. Transport of proteins in and out of the nucleus

In eukaryotic cells, the nucleus is physically separated from the cytoplasm by an impermeable double membrane called the nuclear envelope (NE). The traffic of macromolecules across the NE occurs through nuclear pore complexes (NPCs). NPCs are huge macromolecular assemblies that perforate the nuclear envelope and are responsible for the bidirectional exchange of molecules between the cytoplasm and nucleus through a central channel that has a limiting diameter of ~25–30 nm (Feldherr *et al.*, 2001). NPCs have an estimated mass of ~44 MDa and are constructed from multiple copies of ~30 different proteins collectively called nucleoporins (Cronshaw *et al.*, 2002;Lusk *et al.*, 2004;Stewart, 2007), many of which are conserved between species. Nucleoporins can be divided into three subgroups. The first one is represented by integral membrane proteins that are believed to play a role in NPC assembly and anchoring of the NPC to the membrane. The second group is formed by proteins that contain repeated peptide motives of the type GLFG, FXFG, PSFG or FG and are called FG-nucleoporins. These proteins play a direct role in the transport, and several of them have been shown to interact with karyopherins (Kaps; see below) directly. The third group is represented by the most evolutionary conserved proteins which do not contain the repeated peptide motives. They are thought to provide a scaffold for organization of the FG-nucleoporins (Lusk *et al.*, 2004).

The nuclear transport of small molecules less than 9 nm in diameter occurs via passive diffusion whereas macromolecules greater than 40 kD are transported actively through NPCs

(Pemberton and Paschal, 2005;Stewart, 2007). Transport of most nuclear cargos is accomplished by soluble carrier molecules termed β-Kaps that shuttle between the cytoplasm and nucleus. The nuclear import and export carriers are called importins and exportins. They recognize cargos by binding a nuclear localization signal (NLS) or a nuclear export signal (NES) in either the cytoplasm or the nucleus, respectively, and dock them to the NPC for subsequent translocation (Lusk *et al.*, 2004). The family of β-Kaps includes 14 members in budding yeast and at least 20 in human (Cook *et al.*, 2007). Individual Kaps have the ability to bind specific classes of cargos that provides the basis for the independent regulation of the transport of different classes of molecules (Lusk *et al.*, 2004). While the overall structure of the β-Kaps is believed to be similar, their sequence similarity is low (the identity typically between 15% and 20%), except for a region near the N-terminus that contains a binding site for the GTPase Ran. The differences in the sequences of Kaps are likely related with their ability to recognize different cargos (Pemberton and Paschal, 2005).

In addition to the Kaps, nucleo-cytoplasmic exchange requires the activity of the GTPase Ran (Macara, 2001a; Fried and Kutay, 2003;Weis, 2003). In the nucleus, Ran is maintained in its GTP-bound state by the nuclear-restricted GTP exchange factor, Ran-GEF. In contrast, the Ran GTPase activating protein (Ran-GAP) is primarily cytoplasmic, ensuring that this pool of Ran is in its GDP-bound form. The formation of import complexes between β-Kaps and their cargos is stable in the presence of cytoplasmic Ran-GDP. However, once the β-Kap–cargo complexes enter the nucleus, Ran-GTP binds to the β-Kaps and displaces their cargo. On the other hand, the formation of export complexes is stabilized in the nucleus by Ran-GTP, and as soon as these complexes reach the cytoplasm the GTP is hydrolyzed and the complex disassembles. Moreover, the Ran-GTP gradient provides energy for recycling of Kaps back to the cytoplasm and continued rounds of transport.

Additional factors contribute to the Ran cycle. Ran-GAP needs cofactors (the Ran-binding proteins, or RanBPs) to act on the Kap-bound complexes that reach the cytoplasmic side of the nuclear envelope. Following hydrolysis, Ran-GDP is recycled back to the nucleus by a dedicated transport factor (nuclear transport factor 2 or NTF2) that bears no resemblance to the Kaps (Stewart, 2000).

1.2.1. Nuclear protein import pathways

There are different protein nuclear import pathways that use different carriers, but share many common features and are based on a concerted series of protein-protein interactions by which cargos are recognized in the cytoplasm, translocated through NPCs, and released into the nucleus (Stewart, 2007). In each pathway, cargo proteins are targeted for nuclear import by short nuclear localization signals (NLSs) sequence motifs which are necessary and sufficient to target proteins into the nucleus. There are different NLS classes, each of which is recognized by the components of a different pathway (Pemberton and Paschal, 2005).

The classical nuclear protein import pathway is responsible for the transport of a broad range of cargos and has been studied in substantial biochemical, genetic, cell biological and structural detail (Lange *et al.*, 2007;Stewart, 2007). The classical nuclear protein import cycle generates transport rates of ~100–1,000 cargos per minute per NPC (Ribbeck and Gorlich, 2001). The nuclear protein import cycle can be divided into four steps: assembly of the cargo- import carrier complex in the cytoplasm, translocation through NPCs, disassembly of the import complex in the nucleus, and importin recycling. Cargo proteins with conventional NLSs are imported by the carrier importin β1, which binds them through the adaptor protein importin α (Pemberton and Paschal, 2005;Stewart, 2007) and facilitates their movement through NPCs. FG-nucleoporins are thought to be important for mediating the movement of cargo/carrier complexes through NPCs. The phenylalanine side chains of the hydrophobic FG-repeat cores bind to hydrophobic cavities on the surface of carriers (Cook *et al.*, 2007). The interaction is weak (usually of the order of µM affinity) and so is sufficiently transient to enable rapid transport of cargo/carrier complexes (a high affinity would imply slow off-rates). In the nucleus, Ran-GTP binds to importin β1, dissociating the import complex and releasing the cargo. Importin β1 in complex with Ran-GTP is recycled to the cytoplasm, whereas importin α is exported in complex with the karyopherin CAS and Ran-GTP. Finally, cytoplasmic Ran-GAP stimulates the Ran-GTPase, generating Ran-GDP, which dissociates from the importins and thereby releases them for another import cycle (Stewart, 2007).

Conventional NLSs are divided into three broad classes, and two of them are represented by highly basic NLSs. Monopartite NLSs resembling NLS of the SV40 large tumor antigen (PKKKRKV) (Kalderon *et al.*, 1984), comprise a short stretch of 4–5 basic amino acids, whereas bipartite NLSs, such as that of nucleoplasmin (KRPAATKKAGQAKKKK) (Robbins *et al.*, 1991a), consist of two stretches of basic amino acids separated by a spacer of about 10 amino acids. The conventional NLSs of the third type are represented by charged/polar residues which are interspersed with non-polar residues (as in the NLS of the yeast homeodomain containing protein Mata2), or the basic cluster which is surrounded by the proline and aspartic acid residues (as in the protooncogene c-myc (PAAKRVKLD)) (Dang and Lee, 1988). All above-listed classes of NLS are recognized in the cytoplasm by the heterodymeric importin α/β1 complex (Conti *et al.*, 1998;Conti and Kuriyan, 2000;Fontes *et al.*, 2000). There are at least five isoforms of importin α, and each carrier binds a specific range of cargos (Stewart, 2007). Usually, NLSs have ~10 nM affinity for the importin α/β1 complex (Kutay *et al.*, 1997a;Matsuura *et al.*, 2003a;Goldfarb *et al.*, 2004;Matsuura and Stewart, 2005) and the rate of nuclear import correlates with the strength of binding to importin α (Hodel et al., 2006).

A few cargo proteins bind directly to importin β1 rather than through importin α (Cingolani et al., 2002a;Lee et al., 2003a;Lee et al., 2003c). For example, in the absence of importin α, importin β1 binds targeting sequences in transport substrates such as the T-cell protein tyrosine phosphatase (TCPTP) (Tiganis et al., 1997), human immunodeficiency virus (HIV-1) Rev protein (Truant and Cullen, 1999), sterol regulatory element binding protein 2 (SREBP2) (Nagoshi et al., 1999;Nagoshi and Yoneda, 2001) and parathyroid hormone-related protein (PTHrP) (Lam et al., 1999), docks these proteins at the NPC, and interacts with Ran to mediate translocation into the nucleus. Co-crystal structures of importin β1 with a fragment of Kap α, with a fragment of the transcription factor SREBP2, and with the parathyroid hormone-related cargo protein (PTHrP) showed that distinct contacts are made between the importin β1 and each of its cargos (Cingolani et al., 1999a; Cingolani et al., 2002a; Lee et al., 2003a). This suggests that each Kap can have multiple binding sites and explains how a limited number of Kaps can import diverse cargos with no apparent sequence similarity between cargos. Moreover, these structures show that the Kap is capable of adopting different conformations depending on the cargo (Cingolani et al., 1999a;Cingolani et al., 2002a; Lee et al., 2003a).

A variety of nonconventional NLSs that are devoid of basic residues have been identified. Some of them have been characterized as nonconvential importin α interacting motifs, for example, the influenza virus NP protein (Wang et al., 1997), the cellular transcription factor Stat1 (Melen et al., 2001;McBride et al., 2002) or the NLS of UL84 protein of human cytomegalovirus containing 282 amino acids residues, which all are required for binding to the importin α proteins (Lischka et al., 2003).

Some nonconventional NLSs are transported by Kaps different from importin β1. At that, most Kaps of importin β family bind cargos directly and therefore do not rely on an adapter (Pemberton and Paschal, 2005). In some cases, the NLS contains several basic amino acids as has been determined for core histones, ribosomal proteins and arginine–glycine-rich NLSs observed in some RNA-binding proteins (Pemberton and Paschal, 2005). In other cases, the NLS domain is relatively large, raising the possibility that the three-dimensional structure of the protein is critical (Rosenblum et al., 1998). One of the examples of the nonconventional NLS is the 38 amino acid M9 nuclear targeting (shuttle) sequence which is rich in glycine and aromatic residues. It was first defined for the large heterogeneous human mRNA-binding protein hnRNP A1 (Pollard et al., 1996) and later has been identified in a number of other proteins (Siomi et al., 1997;Nakielny and Dreyfuss, 1999). The M9 sequence is recognized by transportin, which is a close homologue of importin β, and mediates both import into and export out of the nucleus. Shorter NLS motifs that bear no obvious resemblance to the classical NLS sequence have also been described, as for example in Sam68 (P-P-X-X-R) (Ishidate et al., 1997) and Cdc6 (S/T-P-X-K-R-L/I) (Takei et al., 1999) proteins. A glycine-arginine (GR)-motif has also been reported to mediate the nuclear translocation of the large fibroblast growth factor FGF-2 isoforms (Dono et al., 1998). Thus, it seems that in addition to the classical basic-type NLS, a variety of other sequences are also able to mediate the nuclear import of proteins.

1.2.2. Nuclear export pathways

Nuclear export of proteins is closely analogous to the nuclear import and involves specific, but distinct, targeting signals, importin homologs and nucleoporin binding sites, as well as Ran and its modifying factors. Proteins that are exported from the nucleus in the cytosol often possess recognizable stretches of amino acids comprising nuclear export sequences (NES). A leucine-rich motif (Wen et al., 1995; Fischer et al., 1995a) and a glycine-rich motif (Michael et al., 1995) have been shown to function as NESs.

The best characterized NES is the hydrophobic leucine-rich NES containing three to four hydrophobic residues (Wen et al., 1995;Fischer et al., 1995a). This signal is utilized in all eukaryotes, and at least 75 proteins containing leucine-rich NESs have been identified (Pemberton and Paschal, 2005). These include many transcription factors and cell cycle regulators, as well as the viral HIV Rev protein and the protein kinase A inhibitor where the hydrophobic NES was first described (Fischer et al., 1995a; Wen et al., 1995; Pemberton and Paschal, 2005). Leucine-rich NESs are recognized by the karyopherin CRM1 (Chromosome maintenance region 1/Exportin 1/Xpo1p/ Kap124p) (Fornerod et al., 1997;Stade et al., 1997;Ossareh-Nazari et al., 1997;Kudo et al., 1997;Ohno et al., 1998). CRM1 binds to the leucine-rich NES directly and mediates export through the NPC in a manner inhibited by the antibiotic leptomycin B (LMB) (Nishi et al., 1994;Ullman et al., 1997;Kudo et al., 1999;Henderson and Eleftheriou, 2000). Like importin β1, CRM1 can also mediate the export of several cargos via adapter proteins (Pemberton and Paschal, 2005).

In addition to CRM1, three members of karyopherin β family have been identified that function as nuclear export carriers. CAS (Cse1p/Kap109p) exports importin α to regenerate cytoplasmic importin for further cycles of transport, whereas exportin-t (Los1p) is responsible for the export of tRNA from the nucleus (Kutay et al., 1997a; Kutay et al., 1998; Arts et al., 1998). Yeast Kap Msn5p is responsible for the transport of the transcription factor Pho4 (Kaffman et al., 1998). This Kap is unusual in that it can mediate both nuclear import and export (Yoshida and Blobel, 2001).

1.2.3. Regulation of nuclear transport

The nuclear transport of proteins can be regulated in two different ways: either directly affecting Kap binding to the cargo, or regulating the interaction between the Kaps and nucleoporins (Lusk et al., 2004).

1.2.3.1. Regulation of the karyopherin-cargo interaction

A number of specific mechanisms regulate nuclear transport precisely, in response to a variety of signals such as hormones, cytokines and growth factors, cell-cycle signals, developmental signals, immune challenge and stress. Masking or exposing of NLS by binding proteins are the common regulatory pathways of the nuclear migration. These processes are mediated by phosphorylation or dephosphorylation, acetylation, ubiquitination or sumoylation. Posttranslational modification of signalling molecules through phosphorylation/dephosphorylation is the best understood mechanism to regulate nuclear transport (Jans et al., 2000;Pemberton and Paschal, 2005) and can be mediated by many different kinases/phosphatases. Since kinases/phosphatases can be regulated by many different cellular signals, signal-responsive phosphorylation/dephosphorylation represents a direct link between extracellular signals and response in terms of nuclear import or export of specific signalling molecules such as cell-cycle regulators, kinases and transcription factors. Many nuclear proteins possess both NLS and NES, meaning that the precise level of nuclear accumulation can be tightly regulated through the modulation of nuclear import as well as of nuclear export (Stommel et al., 1999;Johnson et al., 1999). Moreover, some proteins contain several NLSs or NESs, and interplay between these signals determines the nuclear concentration of a protein.

Phosphorylation is one of the major mechanisms of nuclear transport regulation (Poon and Jans, 2005). It either enhances the binding of cargos to Kaps or masks an NLS or NES preventing their recognition by Kap and thus avoiding their import or export. The enhancement of nuclear import or export by phosphorylation of sites close to the targeting signal can occur via two mechanisms: either phosphorylation triggers conformational change that reveals the binding site on NLS or NES, or, alternatively, the receptors can recognize only a phosphopeptide. Masking of NLS by nearby phosphorylation also occurs via several mechanisms (Poon and Jans, 2005). First, phosphorylation might disturb electrostatic interaction between highly basic NLSs and Kaps. The phosphogroup can neutralize a positive charge of the NLS and thus prevent interaction with an NLS recognition molecule. Second, the presence of a phosphate may induce changes in conformational structure of the cargo protein thus preventing tight binding to a Kap and consequently inhibiting nuclear import (one of the types of intramolecular masking – see below). Third, in bipartite NLSs, phosphorylation within the intervening spacer may interfere with the nuclear import machinery.

Prevention of the targeting signal recognition might occur via intramolecular masking, intermolecular masking or nuclear/cytoplasmic retention. Intramolecular masking occurs when the accessibility of the NLS/NES is inhibited by the charge or conformation of the NLS/NES-containing protein (Poon and Jans, 2005). For example, the precursor form of the transcription factor nuclear factor kappa B (NF-kB) p50, p105, has a masked and inaccessible NLS. During an immune response, specific phosphorylation and degradation of the p105 C-terminus unmasks the NLS in the p50 form, enabling recognition by the importin α/β1 complex and nuclear accumulation (Riviere et al., 1991;Henkel et al., 1992). Formation of disulfide bond between cysteine residues can also cause the conformational changes of proteins. For example, in response to oxidative stress, NES of yeast transcription factor Yap1p becomes inaccessible for exportin 1 due to the formation of an intramolecular disulfide linkage (Kuge et al., 2001).

Target sequence recognition may also be prevented by intermolecular masking when the binding of a heterologous protein hides an NLS/NES from a corresponding Kap (Poon and Jans, 2005). Thus, in the absence of immune challenge, the NLS of NF-kB p65 is masked from recognition by the importin α/β1 complex due to binding of the specific inhibitor protein I-kB (Henkel et al., 1992;Huxford et al., 1999). Upon immune challenge, I-kB is phosphorylated, leading to its ubiquitin-dependent degradation by the proteosome and subsequent unmasking of the NF-kB p65 NLS (Henkel et al., 1992;Huxford et al., 1999). In response to DNA damage, the tumour-suppressor p53 forms a tetramer that results in masking of the C-terminal NES. The tetramerization domain overlaps NES of p53, and dissociation of the tetramer is necessary to unmask the NES and allows nuclear export (Stommel et al., 1999). NESs/NLSs can also be masked by ligand binding, as shown for the NES of the androgen receptor, which is located in the ligand binding domain of the molecule. In the presence of a ligand (e.g. androgen), the NES is masked and cannot be recognized by CRM1. Re-localization of androgen receptor to the cytoplasm occurs only after dissociation of the ligand (Saporita et al., 2003). Intermolecular masking of targeting signals can also occur via RNA or DNA binding (Poon and Jans, 2005). Protein-protein interactions can not only mask targeting signals but also enhance nuclear import/export, possibly by facilitating recognition of NLS/NES by the corresponding Kaps (Poon and Jans, 2005).

Another mechanism of regulating nuclear transport is through cytoplasmic or nuclear retention, *i.e.* the binding of NLS/NES-containing cargo to specific cytoplasmic or nuclear factors that anchor or retain cargos in cytoplasmic or nuclear compartments. For example, a small protein angiogenin is able passively enter the nucleus, and its NLS does not mediate interaction with importins but confers binding to nuclear/nucleolar components. These components anchor angiogenin in the nucleus, preventing its diffusion into the cytoplasm (Lixin et al., 2001). Nuclear/cytoplasmic anchoring can be also regulated by phosphorylation as implied by the observation that nuclear retention by the IFi16 NLS appears to be enhanced by CK2 site phosphorylation (Briggs et al., 2001).

1.2.3.2. Regulation of transport by the NPC

In addition to mechanisms modulating interactions between Kaps and their cargos, there are accumulating evidences supporting a role of the NPC in changing levels of nuclear transport (for review, see (Lusk et al., 2004)). For example, the size of the NPC channel can be significantly altered in response to changes in cellular physiology. Analysis of the nuclear transport between proliferating and quiescent BALB/c 3T3 cells revealed that the size of the NPC translocation channel was larger in cells that were actively growing (Feldherr and Akin, 1993;Feldherr and Akin, 1994). The molecular mechanisms causing these changes, however, are not known yet. The regulation of a particular transport pathway can be achieved through modifications of the NPCs composition by modulation of the importin expression level. For example, distinct nuclear import pathways are inhibited during poliovirus infection due to the selective degradation of two nucleoporins, Nup153 and p62 (Gustin and Sarnow, 2001). Another example is the differential expression of importin α1 and α2 in a range of human leukaemia lines; more differentiated lines contained higher levels of both importins and less differentiated lines contained only one isoform and at reduced level (Nadler et al., 1997). Lipopolysaccharide, concanavalin A, or phorbol ester/ionomycin treatment led to an increase in importin α expression in normal blood lymphocytes, indicating that importin levels can be regulated in response to cellular signals (Nadler et al., 1997). Marked differences in the level of mRNA expression of different importin α isoforms were observed across different tissues; whereas importin α 1 shows low to medium level expression in most tissues, mouse and human importin α4, α5 and α6 are highly expressed in testis and to a lesser extent in spleen (Prieve et al., 1996;Tsuji et al., 1997).

The transport of different Kaps through the NPC involves their binding to specific nucleoporins via specific binding sites. A common feature of these binding sites is that they are devoid of FG repeats and have affinities for Kaps that are much stronger than the Kap – FG repeat interaction (Ribbeck and Gorlich, 2001;Pyhtila and Rexach, 2003; Matsuura et al., 2003a). These sites have been linked to the regulation of distinct nuclear import pathways. For example, the deletion of the specific binding site in nucleoporin Nup1p reduces 450-fold the binding affinity for Kap95p and has specific effects on Kap95p–Kap60p-mediated import (Pyhtila and Rexach, 2003).

Similarly, the mutation of the Kap60p binding site on Nup2p affects the efficiency of nuclear import (Gilchrist and Rexach, 2003; Matsuura *et al.*, 2003a).

2. The aim of this study

RS1 is critically involved in down-regulation of the Na^+-D-glucose cotransporter SGLT1 in small intestine (Osswald et al., 2005). In LLC-PK$_1$ cells, RS1 inhibits the release of SGLT1 containing vesicles from the *trans*-Golgi network and inhibits transcription of SGLT1 in confluence-dependent manner (Kroiss et al., 2006). Whereas the mechanism of posttranscriptional regulation of SGLT1 by RS1 has been studied extensively, still little is known about the transcriptional regulation of SGLT1 by RS1 as well as about the regulation of RS1 function. Therefore, the goal of this study was to investigate further pathways of the regulation of RS1 function.

In LLC-PK$_1$ cells, one of the levels of regulation of RS1 is represented by its subcellular distribution. RS1 exhibits differential localization in subconfluent *versus* confluent LLC-PK$_1$ cells being present in the nucleus and the cytoplasm of subconfluent cells and in the cytoplasm of confluent cells. Previously, a nuclear localization signal which directs RS1 into the nucleus has been identified in our lab. It was suggested that phosphorylation might serve as a regulatory mechanism of RS1 nuclear translocation during confluence (Leyerer, 2007). However, the exact mechanism of RS1 nuclear transport and the regulation of RS1 localization during confluence including the role of phosphorylation have not been elucidated. Therefore, the major aim of this study was to clarify the mechanisms underlying nuclear transport of RS1 and its regulation. To this end, several aspects had to be investigated. The first objective of this study was to reveal determinants of confluence-dependent nuclear location of RS1. Several agents modulating cell cycle were applied to reveal whether cell cycle is involved in this type of regulation. Second, we aimed at identification of proteins (importin(s) and exportin(s)) involved in the nucleocytoplasmic translocation of RS1. Third, the role of RS1 phosphorylation and nuclear export in the regulation of RS1 nuclear transport was questioned.

In the second part of this thesis, the gene expression profiling of fibroblasts with RS1$^{-/-}$ genotype in comparison with wild-type fibroblasts was performed in an attempt to characterize the target genes of RS1. These dara are preliminary and are presented in Appendix I.

Regulation of RS1 on the protein expression level includes degradation pathways. This type of regulation plays an important role during confluence in LLC-PK$_1$ cells controlling the amount of the protein in subconfluent *versus* confluent cells. Hence, the possible degradation pathways involved in regulation of RS1 protein expression level were examined. The studies of the posttranscriptional regulation of RS1 protein expression are still ongoing and are presented in Appendices II and III.

3. Materials and methods

3.1. Materials

3.1.1. Chemicals

All laboratory chemicals were of p.a. grade and purchased from Sigma-Aldrich (Taufkirchen, Germany), Merck (Darmstadt, Germany), Carl Roth GmbH (Karlsruhe, Germany), Serva (Heidelberg, Germany), Biozym Diagnostik (Hameln, Germany) or AppliChem (Darmstadt, Germany).

3.1.2. Antibodies

The antibodies used in this work are listed in Tables 1 and 2. In some experiments, antibodies which recognize phosphorylated serine 370 were used. Generation of these antibodies is described in this work (3.2.3.).

Table 1. Primary antibodies used in this work. *IB*, immunoblot; *IP*, immunoprecipitation.

Antigen	Species, specification	Application, dilution	Supplier
GFP	Mouse monoclonal, MMS-118P	IB, 1 : 5 000	Covance, Freiburg, Germany
GFP	Rabbit polyclonal, ab290	IP, 1 µl/ml of lysate	Abcam, Cambridge, UK
Importin β1	Rabbit polyclonal, sc-11367	IB, 1 : 1 000	Santa Cruz Biotechnology, Heidelberg, Germany
Importin β2	Goat polyclonal, sc-6914	IB, 1 : 1 000	Santa Cruz Biotechnology
FLAG	Mouse monoclonal, F1804	IB, 1 : 20 000	Sigma-Aldrich

Table 2. Secondary antibodies used in this work. *IB*, immunoblot.

Antibody	Application, dilution	Supplier
Anti-rabbit IgG, HRP-conjugated	IB, 1 : 5 000	Sigma-Aldrich
Anti-mouse IgG, HRP-conjugated	IB, 1 : 5 000	Dianova, Hamburg, Germany

3.1.3. Affinity matrices

For immunoprecipitation, Affi-Prep Protein A gel (#156-0005, Bio-Rad, Hercules, CA) was used. GFP-tagged proteins were purified on μColumns (#130-042-701, Miltenyi Biotec, Bergisch Gladbach, Germany) using μMACS anti-GFP microbeads (#130-091-125, Miltenyi Biotec) by means of μMACS Separation Unit (#130-042-602, Miltenyi Biotec).

3.1.4. DNA and Protein Markers

DNA markers 1kb and 10 kb Ladder and PageRuler Prestained Protein Ladder (MBI Fermentas, St. Leon-Rot, Germany) were used.

3.1.5. Enzymes

Restriction endonucleases (XhoI, PstI, Eco47III, PstI, Acc65I, BglII and BamHI), Pfu DNA polymerase, and T4 DNA ligase were obtained from MBI Fermentas.

3.1.6. Inhibitors and activators

The following inhibitors and activators were used in this work: protease inhibitor cocktail set III (Calbiochem) (the final concentrations of inhibitors were: 1 mM 4-(2-aminoethyl)benzenesulfonyl fluoride hydrochloride, 0.8 μM aprotinin, 50 μM bestatin, 15 μM N-(trans-epoxysuccinyl)-L-leucine-4-guanidineobutylamide, 20 μM leupeptin, 10 μM pepstatin A); 10 μg/ml phosphatase inhibitor cocktail I (P2850, Sigma-Aldrich) containing cantharidin, bromotetramisole, and microcystin; 10 μg/ml phosphatase inhibitor cocktail II (P5726, Sigma-Aldrich) containing sodium ortovanadate, sodium molybdate, sodium tartrate, and imidazole; kinase inhibitors staurosporine (Sigma-Aldrich), Ro 31-8220, UO126 (Calbiochem); nuclear export inhibitor leptomycin B (LMB); PKC activator phorbol 12-myristate 13-acetate (PMA) (Sigma-

Aldrich); proteasome inhibitors MG-132, MG-262; calpain inhibitor calpeptin (Calbiochem); calcium ionophore A23187 (Sigma-Aldrich); calmodulin inhibitor W-13 (Calbiochem); mimosine; nocodazole (Sigma-Aldrich).

3.1.7. Reaction kits

The indicated reaction kits were used according to the manufacturer's instructions: Plasmid Purification Kit (Qiagen, Hilden, Germany); RNeasy Midi Kit (Qiagen); ECL PlusTM Detection Kit (GE Healthcare, Munich, Germany); Pierce ECL Western Blotting Substrate (Pierce, Bonn, Germany).

3.1.8. Peptides

The peptides were selected from the pig RS1 sequence and contained the serine 370. Peptide containing the phosphorylated serine 370 (ELHELLVIpSSKPALENTSC) with COOH terminal cysteine was synthesized by EZBiolab (Dolan Way, USA), and the identical peptide with a nonphosphorylated serine 370 was synthesized by GenScript (Scotch Plains, USA).

3.1.9. Synthetic Oligonucleotides

The oligonucleotides used in this work are listed in the table 3. All oligonucleotides were synthesized by Biomers.net (Ulm, Germany).

Table 3. Oigonucleotides used in this work. Restriction sites or their parts are shown in bold; cohesive ends are underlined.

Oligos	5´-Sequence-3´	Introduced restriction site
FIL1	CG**CTGCAG**GCTGTCAGCCTTCTGTGGAG	
FIL2	GC**GGTACC**TCAATTTTGGGTCCATCTTTCAG	
ALI-1F	p**TCGAG**CTGAGAATCTTTATTTTCAGGGC	XhoI
ALI-1R	pTGGCGCCCTGAAAATAAAGATTCTCAGC	
ALI-2F	pGCCAGCGCTAAAGAAACCGCTGCTGCTAAA	
ALI-2R	pCGAATTTAGCAGCAGCGGTTTCTTTAGCGC	
ALI-3F	p**TTCG**AACGCCAGCACATGGACAGCT**CTGCA**	PstI
ALI-3R	pGAGCTGTCCATGTGCTGGCGTT	

3.1.10. Plasmids and constructs

Plasmids and expression vectors used in this study are listed in 4.

Table 4. Plasmids and constructs used in this work.

Plasmid	Description	Source (Reference)
pEGFP-C1	mammalian expression vector, allows C-terminal fusion to GFP	Clontech, Heidelberg, Germany
GFP-S-tag-TEV	mammalian expression vector on the basis of pEGFP-C1 expressing fusion protein of GFP and S-tag separated by tobacco etch virus (TEV)-protease cleavage site	This work
GFP-S-TAG-TEV-CK2-NS-PKC-PKC	GFP-S-tag-TEV vector expressing nuclear shuttling signal of hRS1 (amino acids 338-402)	This work
GFP-pRS1	mammalian expression vector coding for the fusion protein of pRS1 N-terminally linked to GFP	K. Baumgarten
GFP-pRS1 (Val368Ala)	Mammalian expression vector coding for the fusion protein of GFP C-terminally linked to pRS1 in which valine 368 was mutated to alanine	V. Gorboulev
GFP-pRS1 (Leu366Ala, Val368Ala)	Mammalian expression vector coding for the fusion protein of GFP C-terminally linked to pRS1 in which leucine 366 and valine 368 were mutated to alanines	V. Gorboulev
GFP-pRS1 (Ser370Glu)	Mammalian expression vector coding for the fusion protein of GFP C-terminally linked to pRS1 in which serine 370 was mutated to glutamate	Leyerer, 2007
pHM830	Mammalian expression vector encoding the fusion protein of β-galactosidase C-terminally linked to GFP that allows cloning of desired fragments between β-galactosidase and GFP	(Sorg and Stamminger, 1999)

Plasmid	Description	Source (Reference)
pHM829	Mammalian expression vector encoding fusion protein of β-galactosidase N-terminally linked to GFP	Sorg, Stamminger 1999
βGal-NS-GFP	pHM830 with insertion of the fragment comprising aa 349-369 of pRS1	Leyerer, 2007
βGal-CK2-NS-PKC-PKC-GFP	pHM830 with insertion of the fragment comprising aa 342-402 of pRS1	Leyerer, 2007
βGal-CK2-NS-PKC-PKC (Ser370Ala)-GFP	pHM830 with insertion of the fragment comprising aa 342-402 of pRS1 with the mutation of serine 370 to alanine	Leyerer, 2007
βGal-CK2-NS-PKC-PKC (Ser370Glu)-GFP	pHM830 with insertion of the fragment comprising aa 342-402 of pRS1 with the mutation of serine 370 to glutamate	Leyerer, 2007
βGal-CK2-NS-PKC-PKC (Ile356Gly)-GFP	pHM830 with insertion of the fragment comprising aa 342-402 of pRS1 with the mutation of isoleycine 356 to glycine	V. Gorboulev
βGal-CK2-NS-PKC-PKC (Ile356Gly, Ile369Gly)-GFP	pHM830 with insertion of the fragment comprising aa 342-402 of pRS1 with the mutation of isoleycines 356 and 369 to glycines	V. Gorboulev
βGal-NS-PKC-PKC-GFP	pHM830 with insertion of the fragment comprising aa 349-402 of pRS1	Leyerer, 2007
βGal-CK2-NS-GFP	pHM830 with insertion of the fragment comprising aa 342-369 of pRS1	Leyerer, 2007
βGal-NS-PKC-GFP	pHM830 with insertion of the fragment comprising aa 349-374 of pRS1	Leyerer, 2007

Plasmid	Description	Source (Reference)
βGal-CK2-NS-PKC-GFP	pHM830 with insertion of the fragment comprising aa 342-374 of pRS1	Leyerer, 2007
GFP-CK2-NS-PKC-PKC-βGal	pHM829 with insertion of the fragment comprising aa 342-402 of pRS1	Leyerer, 2007
GFP-CK2-NS-PKC-PKC (+2Tryp)-β-Gal	pHM829 with insertion of the fragment comprising aa 342-402 of pRS1 in which two additional trypsin cleavage sites are introduced (see Results, subsection 4.7)	V. Gorboulev
hRS1-YFP	mammalian expression vector coding for the fusion protein of hRS1 C-terminally linked to YFP	V. Gorboulev
YFP-hRS1	mammalian expression vector coding for the fusion protein of hRS1 N-terminally linked to YFP	V. Gorboulev
YFP-hRS1-FLAG-His$_8$	mammalian expression vector coding for the fusion protein of hRS1 N-terminally linked to YFP and C-terminally to FLAG-His$_8$-tags	V. Gorboulev
FLAG-His$_8$-hRS1-YFP	mammalian expression vector coding for the fusion protein of hRS1 C-terminally linked to YFP and N-terminally to FLAG-His$_8$-tags	V. Gorboulev
GFP-pRS1-FLAG-His$_8$	mammalian expression vector coding for the fusion protein of pRS1 N-terminally linked to GFP and C-terminally to FLAG-His$_8$-tags	V. Gorboulev

As selective agents the following antibiotics were used:

100 µg/ml ampicillin – for constructs on the basis of pHM829 and pHM830 plasmids and for YFP-hRS1, YFP-hRS1-FLAG-His$_8$, FLAG-His$_8$-hRS1-YFP, and hRS1-YFP;

30 µg/ml kanamycin – for constructs on the basis of pEGFP-C1 plasmid.

3.1.11. Bacteria

The bacterial *E.coli* strain DH10B (Grant *et al.*, 1990) was used for selection and amplification of plasmids.

3.1.12. Cell lines

HEK 293 is a human embryonic kidney cell line (Graham *et al.*, 1977). LLC-PK$_1$ cells represent renal epithelial cells derived from porcine kidney (Hull *et al.*, 1976).

3.1.13. Buffers and solutions

All aqueous solutions were prepared with deionised water and generally autoclaved at 120°C for 20 min. Buffer compositions are given in corresponding sections.

3.1.14. Software

Search of the putative NES motifs and calmodulin binding motifs was performed employing Minimotif Miner (Balla *et al.*, 2006). In addition, Calmodulin Target Database was employed to predict potential calmodulin binding motifs (Yap *et al.*, 2000) (http://calcium.uhnres.utoronto.ca/ctdb/ctdb/home.html). Search of the putative PEST sequences was performed with PEST-FIND program (Rogers *et al.*, 1986). Alignment of NS sequences of porcine RS1 and its orthologs was performed with web-based Clustal X (Version 1.83) (www.searchlauncher.bcm.tmc.edu/multi-align/multi-align.html). Densitometric analysis was performed using program Image J (Rasband, W.S., ImageJ, U. S. National Institutes of Health, Bethesda, Maryland, USA, http://rsb.info.nih.gov/ij/, 1997-2005; Abramoff M.D., 2004). The UniProtKB/Swiss-Prot accession numbers are: hRS1, Q92681; pRS1, Q29106; mRS1, Q9ER99; rbRS1, O02665.

3.2. Methods

3.2.1. Molecular biology

3.2.1.1. Mutagenesis

Majority of vectors used in this work was generated by Dr. V. Gorboulev and Dr. M. Leyerer. The vectors GFP-TEV-S-Tag and GFP-TEV-S-Tag-(CK2-NS-PKC-PKC) were prepared in the context of this work.

3.2.1.2. Generation of the GFP-TEV-S-Tag vector

GFP-TEV-S-Tag was generated on the basis of pEGFP-C1 (Clontech). Phosphorylated oligonucleotides were annealed in pairs (ALI-1R and ALI-1F; ALI-2R and ALI-2F; ALI-3R and ALI-3F; Table 3) (3.2.1.4.) and sequentially ligated together to generate a synthetic DNA duplex (3.2.1.10.). The ligation mix was then treated with restriction endonucleases XhoI and PstI to eliminate large ligation fragments (3.2.1.7.). The corresponding fragment (88/80 bp long) was isolated by preparative agarose DNA gel electrophoresis (3.2.1.9.) and ligated with the pEGFP-C1 vector cut with same enzymes. After ligation the mix was desalted (3.2.1.11.), and used for transformation of *E.coli* cells (3.2.1.12). Then the E.coli cells were grown on agarose plates containing 30 µg/ml kanamycin as selective antibiotics. Individual clones were grown in LB medium containing 30 µg/ml kanamycin, the plasmids were isolated in small scale (3.2.1.13.), and the presence of the insert was approved by digestion with the restriction endonuclease Eco47III (3.2.1.7.), which cuts the construct with insertion but not the original plasmid. The sequences of selected clones were confirmed by DNA sequencing. DNA of approved constructs was isolated in large scale (3.2.1.13.). Concentration of DNA was measured by spectrophotometry (3.2.1.14.) and adjusted to 1 µg/ml.

3.2.1.3. Generation of the GFP-TEV-S-Tag-CK2-NS-PKC-PKC vector

The fragment of hRS1 containing amino acids 338-403 was amplified by polymerase chain reaction (PCR) (3.2.1.5.). The efficiency of the polymerase chain reaction was verified by analytical agarose gel electrophoresis (3.2.1.8.). Subsequently, the resulting DNA fragment was digested with restriction endonucleases PstI and Acc651 (3.2.1.7.), purified by preparative agarose DNA gel electrophoresis (3.2.1.9.), and ligated with GFP-TEV-S-Tag vector that was cut with the same enzymes (3.2.1.10.). After ligation the mix was desalted (3.2.1.11.), and used for transformation of E.coli cells (3.2.1.12). Then the E.coli cells were grown on agarose plates containing 30 µg/ml kanamycin as selective antibiotics. Individual clones were grown in LB medium containing 30 µg/ml kanamycin, the plasmids were isolated in small scale (3.2.1.13.), and the presence of the insert was approved by restriction analysis with restriction endonucleases BglII and BamHI (3.2.1.7.). The sequences of selected clones were confirmed by DNA sequencing. DNA of approved constructs was isolated in large scale (3.2.1.13.). Concentration of DNA was measured by spectrophotometry (3.2.1.14.) and adjusted to 1 µg/ml.

3.2.1.4. Annealing of oligonucleotides

2.25 nmole of each oligonucleotide were incubated in annealing buffer (10 mM Tris-HCl, pH 7.5, 0.1 M NaCl, 1 mM EDTA) at 65^0C for 10 minutes, cooled to room temperature, and stored at -20^0C. To purify the annealed oligonucleotides, 50 µl of duplex oligonucleotides were incubated with 5.5 µl of 1M $MgCl_2$ and 278 µl of ethanol at -20^0C for 20 min. Subsequently, DNA was precipitated by centrifugation at 15 000 g at RT for 5 min and washed with 70% ethanol. The oligonucleotides were diluted in 22.5 µl of Milli Q water and used for ligation.

3.2.1.5. Polymerase chain reaction (PCR)

The primer sequences were complementary to the template DNA and contained flanking PstI and Acc65I restriction sites (Table 3). PCR reaction was performed in a volume of 50 µl in reaction buffer using 10 ng of template DNA (pcDNA3-hRS1), 5 pmol of forward and reverse oligonucleotide primers (FIL1 and FIL2), and 12.5 nmol of dNTPs. Before the reaction initiation, the reaction mix was covered with paraffin oil. After heating to 94^0C, 1.25 units of Pfu DNA polymerase were added to initiate the PCR reaction. Following an initial denaturation step for 1 min at 94^0C, 25 cycles with the following parameters were performed: denaturation for 30 s at 94^0C, annealing for 30 s at 55^0C, and elongation for 30 s at 72^0C. Elongation time was 1 minute per kb to be amplified. The reaction efficiency was controlled by analytical agarose gel electrophoresis. Prior to restriction digestion, paraffin oil, and polymerase were removed with chloroform extraction.

3.2.1.6. DNA isolation by phenol/chloroform extraction

To precipitate DNA after PCR and remove paraffin oil and proteins, the chloroform extraction was performed. 50 µl of chloroform-isoamyl alcohol (24:1 (v/v)) was added to PCR mix, mixed gently, and two phases were separated by centrifugation at 14 000 g for 10 min (Wallace, 1987). The upper (aqueous) phase containing DNA was used for precipitation of DNA by addition of 0.1 volume of 3M sodium acetate, pH 5.0 and 2.5 volumes of absolute ethanol for 2 h at -20^{0}C followed by centrifugation at 14 000 g for 30 min. The pellet was washed with 70% (v/v) ethanol to remove excess of the salt, air-dried and dissolved in water.

3.2.1.7. Digestion of DNA with restriction endonucleases

1µg DNA was used for a digestion reaction with 5-15 units of enzyme in 20 µl of an appropriate buffer. The enzymatic reaction was performed for 3 h at 37^{0}C. The following restriction endonucleases were used: XhoI, PstI, Eco47III, PstI, Acc65I, BglII and BamHI.

3.2.1.8. Analytical agarose gel electrophoresis of DNA

Samples containing 0.1 µg DNA were mixed with gel-loading buffer and subjected on horizontal 1% (w/v) agarose gel containing 0.3 µg/ml ethidium bromide. The electrophoresis was performed for 1 h in TAE buffer at a voltage of 5 V/cm. The DNA molecular weight markers, 1kb and 10 kb Ladder (MBI Fermentas), were used as a reference for size determination of the DNA fragments. The DNA bands were visualized by illumination in UV light (254 nm) using a Dual Intensity Ultraviolet Transilluminator and photo-documented.

TAE buffer	40 mM Tris-acetate, pH 8.0, 1 mM EDTA
Gel-loading buffer (final concentration)	7.5% (v/v) glycerin, 0.06% (w/v) bromphenol blue

3.2.1.9. Preparative agarose gel electrophoresis

The isolation of DNA fragments after restriction was performed using the preparative agarose gel electrophoresis. DNA was combined with loading buffer and applied to 1% low melting point agarose gel. The electrophoresis was performed for 1 h in TAE buffer at a voltage of 5 V/cm. After electrophoresis, the band of interest was visualized as described above (3.2.1.8.) and excised from the gel with a sterile scalpel. The excised agarose block was incubated at 70^{0}C for 5 min, then immediately placed at -70^{0}C for 10 min and subsequently incubated at 37^{0}C till thawing. Followed by centrifugation at 13 000 g for 5 min. The supernatant containing DNA was used for subsequent analysis.

3.2.1.10. Ligation

T4 DNA ligase was used for cloning the DNA fragments into the plasmids and for ligation of duplex oligonucleotides during the generation of the GFP-TEV-S-Tag vector. Ligation reactions were performed in a total volume of 20 µl. The molar ratios vector:insert and duplex:duplex were 1:5 and 1:1, respectively. T4 DNA ligation buffer and 5 units of T4 DNA ligase were added to the mixture and incubated overnight at +14^{0}C. To inactivate the ligase, the reaction mixture was incubated at 70^{0}C for 10-15 min.

3.2.1.11. Analytical agarose gel electrophoresis of RNA

1 µl of RNA sample was diluted in DMSO/glyoxal solution, incubated for 60 min at 50^{0}C and mixed with 2 µl of the gel loading buffer (Ambion, Darmstadt, Germany). The samples were subjected on horizontal 1% (w/v) agarose gel containing 10 mM iodacetate. The electrophoresis was performed for 2 h in BES buffer at a voltage of 40 V/gel. To avoid the formation of a pH-gradient between the catode and anode, the running buffer (BES) was circulated from anode to cathode by a peristaltic pump. The RNA bands were visualized by illumination in UV light (254 nm) using a Dual Intensity Ultraviolet Transilluminator and photo-documented.

BES buffer	10 mM BES, 0.1 mM EDTA, pH 6.7
1x DMSO/glyoxal solution	50 µg/ml Ethidiumbromide, 50% DMSO, 1M Glyoxal in BES buffer

3.2.1.12. Desalting of DNA samples

For electroporation, DNA samples have to be desalted (Dower et al., 1988). First, the phenol extraction of DNA after ligation reaction was performed as described above. To precipitate DNA, a linear polyacrylamide was used as a carrier. 0.1 volume of 3 M sodium acetate, pH 5.2, 12.5 µg linear polyacrylamide, and 2 volumes of ethanol were added to an aqueous phase and DNA was spun down by the centrifugation in a microfuge at 18 000 g for 30 min at RT. The precipitate was then washed with 1 ml of 70% (v/v) ethanol. In order to remove the residual salt traces from the pellet, precipitated DNA was incubated with 500 µl of 70% (v/v) ethanol for 30 min at RT. After centrifugation for 5 min at 18 000 g at RT, DNA was air-dried and dissolved in 10µl of water.

3.2.1.13. Transformation of bacteria and clone selection

Plasmid DNA was introduced into bacteria by the electroporation method (Dower et al., 1988). 20 µl of electrocompetent *E.coli* cells were carefully thawed on ice, mixed with 10 ng plasmid DNA or 1 µl of a desalted ligation mix, and transferred into the pre-chilled electroporation cuvette. After electrical pulse (1.6 kV, 5 ms, Biojet MJ) bacteria were suspended with 1 ml of SOC medium and incubated for 1 h at 37^0C to express the antibiotic resistance conveyed by the plasmid. 250-500 µl of the bacterial suspension were plated on an agar plate containing the corresponding antibiotic and incubated overnight at 37^0C. After 16 h the single colonies of transformed bacteria were observed, and 10 colonies were selected for further analysis. The cells were transferred to a new agar plate and into tubes with 3 ml LB medium and incubated for 16 h at 37^0C.

SOC medium	10g/l yeast extract, 20g/l bacto-tryptone, 10mM NaCl, 2.5 mM KCl, 10mM $MgCl_2$, 10mM $MgSO_4$, 20mM D-glucose
LB medium	2 % (w/v) bacto-tryptone, 0.5% (w/v) yeast extract, 170 mM NaCl
LB-agar	LB-media, 1.5 % (w/v) agar

3.2.1.14. Isolation of plasmid DNA from E.coli

Plasmids were isolated from overnight cultures inoculated with a single colony. Cultures were grown at 37^0C in LB medium containing the corresponding antibiotic.

Small scale isolation of plasmid DNA (Alkaline lysis, (Birnboim and Doly, 1979))

Cells from 3 ml of the overnight culture were transferred to a snap vial (Eppendorf). After centrifugation for 10 min at 5 000 g, the pellet was resuspended in 300 µl of P1 buffer and incubated for 5-10 min at RT. After addition of 300 µl of P2 buffer tubes were gently mixed and incubated on ice for 5 min. 300 µl of P3 buffer were added and the samples were incubated for 15 min on ice. The cell debris and chromosomal DNA were spun down by the centrifugation for 20 min at 15 000 g at RT, and plasmid DNA was precipitated from 800 µl of supernatant with 640 µl of isopropanol and spun for 30 min at 15 000 g at RT. The pellet was washed with ice-cold 70% (v/v) ethanol, air-dried and resuspended in 20 µl of water. 1 µl of DNA solution was used for an analytical restriction digest. The positive samples were subjected to sequencing. DNA samples were stored at -20^0C.

P1 buffer	50 mM Tris-HCl, pH 8.0, 10 mM EDTA, 100 µg/ml RNase A
P2 buffer	200 mM NaOH, 1% (w/v) SDS
P3 buffer	10 mM Tris-HCl, pH 8.0, 1 mM EDTA

Large scale isolation of plasmid DNA

For large scale purification of plasmid DNA, 100 ml of an overnight *E.coli* culture transformed with the plasmid of interest were used. The purification was performed according to the manufacturer's instructions (HiSpeed Midi Kit, Qiagen, Hilden, Germany). The DNA concentration was determined by spectrophotometry, adjusted to 1 µg/µl concentration, and DNA was stored at -20^0C.

3.2.1.15. Determination of DNA and RNA concentration by spectrophotometry

The DNA concentration was determined by measuring the optical density of a sample at a wavelength of 260 nm. For calculations, extinction coefficients of 0.02 $(µg/ml)^{-1}cm^{-1}$ and 0.025 $(µg/ml)^{-1}cm^{-1}$ were used for a double stranded DNA and RNA, respectively. Purity of DNA was estimated from the ratio of absorbancies at 260 nm and 280 nm. The sample was considered to be free of protein contamination if the ratio was 1.8 - 2.0.

3.2.2. Protein analysis methods

3.2.2.1. Preparation of the whole-cell extracts

Cells were washed with ice-cold PBS and disrupted in the lysis buffer (75 mM Hepes, pH 7.5, 1.5 mM EGTA, 1.5 mM $MgCl_2$, 150 mM KCl, 15% (v/v) Glycerol, 0.075% (v/v) Igepal CA-630) supplemented with protease inhibitor cocktail set III (Calbiochem) and disruptedby sonication (3 min of total sonication time with repeating on and off steps for 15 s and 45 s, respectively). Lysates were centrifuged at 100 000 g for 1 h at 4^0C, and the supernatants were stored at -80^0C.

3.2.2.2. Purification of GFP fusion proteins

Purification of GFP fusion proteins was performed using µMACS magnetic anti-GFP beads (Miltenyi Biotec). This method was used to purify the following proteins: GFP-CK2-NS-PKC-PKC-β-Gal, GFP-CK2-NS-PKC-PKC(+2Tryp)-β-Gal and YFP-hRS1. For purification of GFP-

CK2-NS-PKC-PKC-β-Gal, preconfluent or 30% confluent LLC-PK$_1$ cells were transfected with the corresponding construct, and two days after transfection purification was performed. The two samples were referred as purified from subconfluent and confluent cells. The reason for the use of preconfluent cells for transfection was the low efficiency of transfection of LLC- PK$_1$ cells grown two days after confluence which was not sufficient for the effective purification.

All solutions were pre-cooled at +4^{0}C and all experiments were performed on ice. To avoid the degradation of the proteins, all buffers were supplemented with protease inhibitor cocktail set III (Calbiochem). For study of RS1 phosphorylation state, phosphatase and kinase inhibitors were added as well. To inhibit phosphatase activity, 10 µg/ml phosphatase inhibitor cocktail 1 for serine/threonine protein phosphatases and L-isozymes of alkaline phosphatases (Sigma-Aldrich) and 10 µg/ml phosphatase inhibitor cocktail 2 for tyrosine phosphatases, acid and alkaline phosphatases (Sigma-Aldrich) were used. Compositions of inhibitor sets are desribed in Materials section (subsection 3.1.6). To inhibit kinase activity, following kinase inhibitors were added: 100 nM Ro-31-8220, 200 nM staurosporine, 700 nM UO126, 0.1 µM calphostin, 0.1 mM 5,6-Dichlorobenzimidazole riboside (DRB). For study of RS1 ubiquitination state, N-ethylmaleimide (NEM) was added to inhibit deubiquitinating enzymes.

LLC-PK$_1$ or HEK 293 cells expressing a GFP fusion protein were lysed in the lysis buffer (0.05M Tris-HCl, pH 8.0, 0.5M NaCl, 1% (v/v) Igepal CA-630) as described above (3.2.2.1.). Supernatant was incubated overnight with anti-GFP antibodies coupled to µMACS superparamagnetic microbeads. After incubation, a MACS µ-column was placed in the magnetic field of a µMACS separation device and the suspension of the lysate with microbeads was loaded on the µ-column. The beads were washed 5 times with excess of the lysis buffer followed by washing with 20 mM Tris-HCl, pH 7.5, and fusion protein was eluted by pH shift with the elution buffer (0.1 M triethylamine, pH 11.8, 0.1% (v/v) Triton X-100). Eluate was collected in a tube containing 1M MES, pH 3.0 for neutralization. The volume of the elution fraction was 50 µl; protein concentration was estimated as 0.1-0.2 µg/µl according to SDS-polyacrylamide gel electrophoresis and Coomassie staining.

3.2.2.3. Immunoprecipitation of GFP fusion proteins and associated proteins

All buffers were pre-cooled at +4^{0}C and all experiments were performed on ice. To avoid the degradation of the proteins, protease inhibitor cocktail set III (Calbiochem) was added to all buffers. For inhibition of phosphatase activity, 10 µg/ml phosphatase inhibitor cocktail 1 for serine/threonine protein phosphatases and L-isozymes of alkaline phosphatases (Sigma-Aldrich) and 10 µg/ml phosphatase inhibitor cocktail 2 for tyrosine phosphatases, acid and alkaline phosphatases (Sigma-Aldrich) were used. For compositions of inhibitor sets, see section 3.1.6. For inhibition of kinase activity, following kinase inhibitors were added: 100 nM Ro-31-8220, 200 nM staurosporine, 700 nM UO126, 0.1 µM calphostin, 0.1 mM DRB.

Stably transfected HEK cells expressing GFP-S-tag-TEV or GFP-S- tag -TEV-CK2-NS-PKC-PKC (hRS1) were lysed in the lysis buffer (75 mM Hepes, pH 7.5, 1.5 mM EGTA, 1.5 mM $MgCl_2$, 150 mM KCl, 15% (v/v) Glycerol, 0.075% (v/v) Igepal CA-630) as described above (3.2.2.1.). Cells were lysed by sonication (3 min of total sonication time with repeating on and off steps for 15 s and 45 s, respectively) followed by centrifugation at 4^0C at 100 000 g for 1 h. 2 µl of rabbit polyclonal anti-GFP antibody (Abcam) were added to 2 ml of the supernatant containing 5 mg of total protein and incubated overnight at 4^0C. For precipitation of GFP fusion protein and associated proteins, 100 µl of protein A-coupled sepharose beads (Bio-Rad) were added and the mixture was incubated for 1h at 4^0C. The beads were washed 5 times with 10 volumes of the lysis buffer, and GFP fusion proteins with associated proteins were eluted by the pH shift with elution buffer (0.1 M triethylamine, pH 11.8, 0.1% (v/v) Triton X-100). Eluate was collected in a tube containing 1M MES, pH 3.0 for neutralization. The associated proteins were visualized by Western blot analysis (3.2.2.6.).

3.2.2.4. Determination of protein concentration

The protein content of the samples was determined according to Bradford using bovine serum albumin as a standard (Bradford, 1976). 3 µl of protein solution was diluted in 97 µl of water and 900 µl of Bradford reagent (Bio-Rad) and incubated for 5 min at RT. The extinction of the samples was measured at 595 nm and correlated with the extinction of the solvent and control BSA samples (2, 4, 6, 8, and 10 µg).

3.2.2.5. SDS-polyacrylamide gel electrophoresis

Proteins or cell extracts were separated by the discontinuous SDS-polyacylamide gel electrophoresis according to Laemmli (Laemmli, 1970). The gels were composed of two layers, the separating gel containing the corresponding amount (see below) of acrylamide/bisacrylamide, 375 mM Tris-HCl, pH 8.8, and 0.1% SDS and a stacking gel (5% acrylamide/bisacrylamide, 0.1% SDS, 125 mM Tris-HCl, pH 6.8). Depending on the required separation range the acrylamide concentration was adjusted to 10%, 12.5%, 15%, or 17.5% in separating gel buffer using the Rotiphorese gel, acrylamide/bisacrylamide 37.5:1 mixture (Carl Roth GmbH). Shortly before casting, polymerization of the stacking or separating gels was initiated by addition of ammonium persulfate (APS) and N,N,N',N'-Tetramethylethylendiamine (TEMED) to a final concentration of 0.01% (v/v) of each component.

Protein samples were prepared by heating for 5 min at 95^0C in SDS sample buffer (0.001 % (w/v) bromphenol blue, 10 % (v/v) glycerole, 0.25 M β-mercaptoethanol, 1 % (w/v) SDS, 15 mM Tris-HCl, pH 6.8). The samples were loaded onto polyacrylamide gel using Gel-Loadertips (Hartenstein). The electrophoresis was performed in SDS running buffer (24.8 mM Tris-HCl, pH 8.3, 192 mM glycine, 0.1% (w/v) SDS) at 25 V/cm using MiniProtean-3 electrophoresis chambers (Biorad) and the electrophoresis power supply EPS601 (GE Healthcare). The PageRuler Prestained Protein Ladder (MBI Fermentas) was used as a size reference.

3.2.2.6. Western blot and immunodetection

For immunodetection of proteins, the samples were subjected to immunoblot analysis. Proteins were separated by SDS-PAGE, transferred onto a Polyvinylidene Difluoride (PVDF) membrane using a semi-dry system (Gershoni and Palade, 1983), and a protein of interest was detected by antibodies. For semi-dry transfer of proteins the horizontal semi-dry transferblot (chamber type SD 18) with two graphite plates was used. Before transfer, the PVDF membrane was pre-soaked in methanol for 5 min. Thereafter, Whatman filter papers, the gel and the membrane were incubated in the blotting buffer (25 mM Tris, pH 8.3, 192 mM glycine, 10 % (w/v) methanol) for 10 min. Subsequently, the sheets of Whatman paper, the PVDF membrane, and the gel were assembled on a graphite plate of the transferblot in the following order: (cathode), 3 x Whatman filter paper, SDS-PAGE gel, PVDF membrane, 3 x Whatman filter paper, (anode). Blotting was performed in the blotting buffer at 1.5-2 mA/cm^2 for 2 h. After blotting, the PVDF membrane was stained with Ponceau dye (2% (w/v) Ponceau S, 1% (v/v) acetic acid) (Salinovich and Montelaro, 1986) to detect the presence of proteins. The membrane was incubated for 40 s in the dye and washed thoroughly with water. The dye was washed away by the next washing step with TBST buffer (137 mM NaCl, 50 mM Tris-HCl, pH 8.0, 2.7 mM KCl, 0.05 % (v/v) Tween 20).

Immunodetection of the proteins was accomplished as following. First, all non-specific binding sites on the membrane were blocked by incubation with blocking buffer (5 % (w/v) milk powder in TBST buffer) for 1 h at RT. Then the membrane was incubated with the primary antibodies diluted in TBST containing 1% (w/v) milk for 1 h at RT or overnight at +4^{0}C. After three 5 min washing steps with TBST containing 1% (w/v) milk, the membranes were incubated with the horse radish peroxidase (HRP)-coupled secondary antibodies diluted in TBST containing 1% (w/v) milk for 1 h at RT. Unbound antibodies were removed with next three 5 min washing steps with TBST. During all incubation steps the membrane was incubated on the rotator. The bound label was visualized by enhanced chemiluminescence using ECL Plus Detection Kit (GE Healthcare) or Pierce ECL Western Blotting Substrate (Pierce) according to the manufacturer's instructions. The obtained pictures were scanned and densitometric analysis was performed using program Image J (http://rsbweb.nih.gov/ij/index.html).

For investigation of the phosphorylation status of serine 370 using phosphospecific antibodies, the immunodetection procedure was modified as following. First, all non-specific binding sites on the membrane were blocked by incubation with blocking buffer (3% (w/v) BSA in PBS buffer) for 1 h at RT. Then the membrane was incubated with the primary antibodies diluted in the blocking buffer for 1 h at RT or overnight at +4^{0}C. After three 5 min washing steps with PBS, the membranes were incubated with the horse radish peroxidase (HRP) - coupled secondary antibodies diluted in PBS for 1 h at RT. Unbound antibodies were removed with next three 5 min washing steps with PBS. The detection and analysis of the bands was accomplished as described above.

Bound antibodies were removed from the Western Blot membranes by incubation in the stripping buffer containing 0.1M glycine, pH 2.9 for 1 h at RT. Afterwards membranes were washed three times in TBST and blocked with the blocking buffer for 1 h. After this procedure Western blots could be used for the re-probing with other antibodies.

3.2.2.7. Staining of protein polyacrylamide gels with Coomassie blue

Detection of proteins separated by SDS-PAGE was accomplished by staining the gels with Coomassie Brilliant blue R250 (Merck). The gels were incubated in Coomassie staining solution (0.1 % (w/v) Coomassie brilliant blue R250, 7.5 % (v/v) acetic acid, 50 % (v/v) methanol) with gentle agitation for 30 min. In order to remove nonbound dye from the protein gels, the gel was subsequently destained by three 10 min incubations in the destaining solution (40 % (v/v) methanol, 10 % (v/v) acetic acid). Alternatively, gels were destained by repeated boiling in water using a microwave. Moist gels were kept in 12% (v/v) acetic acid at 4^{0}C in sealed plastic bags or subjected to gel drying.

3.2.2.8. Silver staining of protein polyacrylamide gels

Silver staining methods are about 10-100 times more sensitive than Coomassie Blue staining techniques. Thus, they were used for detection of very low amounts of protein on electrophoresis gels. Protein detection is based on the binding of silver ions to the amino acid side chains, primary the sulfhydril and carboxyl groups of proteins (Switzer *et al.*, 1979;Oakley *et al.*, 1980;Merril *et al.*,

1981;Merril and Pratt, 1986), followed by reduction to free metallic silver (Rabilloud, 1990). The protein bands are visualized as spots where the reduction occurs and, as a result, the image of protein distribution within the gel is based on the difference in oxidation-reduction potential between the gel's area occupied by proteins and the free adjacent sites.

After electrophoresis, the gel was soaked in fixing solution for 2 h. Fixation restricts protein diffusion from the gel matrix, removes interfering ions and detergent from the gel and improves the sensitivity of the staining, and decreases the background. The fixing solution was discarded, and the gel was washed two times with 50% (v/v) ethanol for 30 min to remove the remaining detergent ions as well as fixation acid from the gel. Then the gel was incubated in sensitizing solution for 2 min with gentle rotation. This step increases the sensitivity and the contrast of the staining. The gel was then washed twice, 1 min each time, with water. The staining was performed by incubation with the cold silver staining solution for 20 min that allowed the silver ions to bind to proteins. Afterwards the gel was rinsed twice with water to remove the excess of unbound silver ions. The protein image was developed by incubation of the gel in the developing solution for 2 - 10 min. The reaction was stopped by addition of the terminating solution as soon as the desired intensity of the bands was reached. Moist gels were kept in 12% (v/v) acetic acid at 4^0C in sealed plastic bags or subjected to gel drying.

Fixing solution	50% ethanol, 12% acetic acid, 0.05% formaldehyde
Sensitizing solution	0.02% $Na_2S_2O_3$ x 5 H_2O
Silver staining solution	0.1% $AgNO_3$, 0.07% formaldehyde
Developing solution	3% Na_2CO_3, 0.05% formaldehyde, 0.0002% $Na_2S_2O_3$ x 5 H_2O
Terminating solution	1% glycine

3.2.2.9. Gel drying

Polyacrylamide protein gels were soaked for 0.5-2 h in 10% (v/v) glycerol, covered with cellophane and dried using drying frames for 1-2 days.

3.2.3. Generation and testing of phosphospecific antibodies

Generation of rabbit polyclonal antibodies which recognise serine 370 specifically in phosphorylated form included several stages. First, the peptide containing phosphorylated serine 370 (purity >95%, 30-39 mg) and the corresponding non-phosphorylated peptide (>70%, 30-39 mg) were synthesized. The peptides were used for immunisation, ELISA titration and affinity purification. For immunisation of rabbits, the phosphorylated peptide was conjugated to a carrier protein ovalbumin and injected into the host animals (rabbits). Two animals were used for each immunisation in view of the variability of the host responses to the antigen. Then, the antisera were tested with Enzyme-linked Immunosorbent Assay (ELISA). The bleed of a better responding host (the antiserum which had the highest titer and a relatively low cross-reactivity against non-phosphopeptide) was used for the isolation of the antibodies by means of sequential nonphospho- and phosphoaffinity purifications. Antibodies recognising the phosphorylated epitope were separated from the nonphospho-specific antibodies using the non-phosphorylated peptide column. The purpose of the second purification using the phosphoaffinity column was to obtain exclusively phospho-specific antibodies. The retained elution fraction should contain the desired phospho-specific antibodies, whilst the flow-through should include antibodies to other epitopes. ELISA was performed against both peptides to test whether the isolated antibodies are specific to the phosphorylation site of interest.

3.2.3.1. Immunization of rabbits

The 19-amino acid peptide, ELHELLVIpSSKPALENTSC, containing serine 370 (PKC consensus phosphorylation site) with COOH terminal cysteine was synthesized by EZBiolab (Dolan Way, USA). The peptide was selected from the pig RS1 sequence and corresponded to amino acids 363–379 of pRS1. The corresponding peptide with a nonphosphorylated serine 370, ELHELLVISSKPALENTSC, was synthesized by GenScript (Scotch Plains, USA) and used for screening and purification as detailed below. The phosphorylated peptide was conjugated via cysteine to the carrier protein ovalbimin using 3-maleinidobensoic acid N-hydroxysuccinate (Sigma). Ovalbumin-coupled peptides were used to produce polyclonal antibodies (Poppe et al., 1997). Six month old rabbits were immunized subscapularly with 1 mg of the ovalbumin-conjugated peptide in Complete Freund's Adjuvant, followed by subsequent subscapularly booster injections with 1 mg of the peptide in Incomplete Freund's Adjuvant every 3 weeks. Sera were obtained from the animals every 3 weeks starting on the day 35 after the initial immunization. 5-20 ml of blood from the ear artery were collected each time. After incubation for 3 h at RT and for 16 h at +4^{0}C, the blood was centrifuged at 10 000 g for 10 min at +4^{0}C, and the supernatant was isolated. The serum was aliquoted in 0.5-1 ml fractions and stored at -70^{0}C.

3.2.3.2. Identification of the antibody titer. Enzyme-linked Immunosorbent Assay (ELISA).

The sera and antibody titers as well as cross-reaction were determined by ELISA employing the phosphorylated or the non-phosphorylated peptides as antigens. The wells of a 96 well PVC microtiter plate (NUNC) were coated with the antigen by pipeting 100 µl of coating buffer containing 1 µg of a peptide per well and incubated for 16 h at +4°C. After removal of the coating solution, the remaining protein-binding sites in the coated wells were blocked by incubation with the blocking buffer containing 0.5 % BSA for 2 h at RT. The plate was washed three times with the washing buffer. The antiserum or the antibodies were diluted in the blocking buffer (progressive dilutions from 1 : 400 to 1 : ~400 000), and 100 µl of the antibodies or the antiserum mixture were added to the wells. The blocking buffer alone served as a negative control. After 16 h incubation at +4°C, the plate was washed three times with the washing buffer. Each well was filled with 100 µl of the blocking buffer containing the goat anti-rabbit secondary antibodies coupled with F(ab) alkaline phosphate in a 1 : 250 dilution and incubated for 4 h at RT protected from light. The secondary antibodies were washed away by three times washing with the washing buffer, and 100 µl of the equilibration buffer were placed in each well for 5 min at RT. The equilibration buffer was decanted and each well was filled with 100 µl of the substrate buffer and incubated for 1 h at RT. The colour reaction was blocked by the stop solution (100 mM EDTA, pH 8.0). The absorption at 405 nm was measured and the absorption at 450 nm (background) was subtracted to give final results.

Coating buffer	0.1 M $NaHCO_3$ titerd to pH 9.6 with 0.1 M Na_2CO_3
Equilibration buffer	150 mM NaCl, 10 mM Tris-HCl, pH 9.8
Blocking buffer	PBS buffer, 200 mM NaCl, 0.5 % BSA, 0.1 % sodium azide
Substrate buffer	150 mM NaCl, 10 mM Tris-HCl, pH 9.8, 5 mM $MgCl_2$, 1 mg/ml para-nitrophenylphosphate
Washing buffer	PBS buffer, 200 mM NaCl, 0.05 % (v/v) Tween 20
Stop solution	100 mM EDTA, pH 8.0
PBS buffer	137 mM NaCl, 20 mM, 2.7 mM KCl, 1.5 mM KH_2PO_4, 8 mM Na_2HPO_4 x $2H_2O$ pH 7.14

3.2.3.3. Affinity purification of antibodies

For the affinity purification, the peptides were immobilized on SulfoLink Coupling Gels (Pierce). The affinity column coupled with non-phosphopeptide was washed with 6 ml of PBS, and 1.6 ml of an antiserum was loaded onto the affinity column and incubated for 1.5 h at RT. The flow-through was collected and applied onto the phosphopeptide-coupled column that was pre-washed with 6 ml of PBS. After incubation for 1.5 h at RT, the column was washed with 16 ml of PBS. The antibodies were eluted by pH-shift with glycine buffer (100mM glycine-HCl, pH 2.5, 0.1% Triton X-100, 0.15 M NaCl). Sixteen fractions of 500 µl each were collected into tubes containing 25 µl of 1M Tris, pH 9.5 and 500 µl of glycerin.

For the regeneration of the column, the column was washed with 16 ml of PBS and 5 ml of 0.05% NaN_3. The columns were stored in 2 ml of 0.05% NaN_3 at +4^{0}C.

3.2.4. Cell Culture

3.2.4.1. Cultivation of mammalian cells

Native and transiently transfected human embryonic kidney HEK 293 cells and porcine renal epithelial LLC-PK_1 cells were grown at 37^{0}C in a humidified 5% CO_2 atmosphere in Dulbecco's modified Eagle's medium (DMEM, Sigma-Aldrich) containing 10% (v/v) fetal calf serum (FCS, Sigma-Aldrich), 1% L-glutamine (PAA, Pasching, Austria), and 1% penicillin/streptomycin (PAA). In case of stably transfected cells, the medium was supplemented with 0.8 mg/ml G418. Cells were grown to a maximal density of 1-2 x10^6 cells/cm^2 and passaged every 2-3 days.

3.2.4.2. Passage

Serial passages were done 2-3 times a week. For passage, HEK 293 cells were detached mechanically whereas LLC-PK_1 cells were detached by 5-10 min incubation in detaching buffer (Ca^{2+}- and Mg^{2+}-free Dulbecco's phosphate buffered saline (PBS) (137 mM NaCl, 2.7 mM KCl, 1.5 mM KH_2PO_4, 8 mM Na_2HPO_4) (Sigma-Aldrich) supplemented with 28 mM $NaHCO_3$, 0.5 mM EDTA, and 10 mM HEPES) until all cells were detached from the culture dish. The aspirated cells were pelleted by 10 min centrifugation at 1 000 g and resuspended in the cultivation medium. 20-30 % of the cells were transferred into new flask.

3.2.4.3. Cryoculture

Detached and sedimented cells were resuspended in cryomedium (DMEM, 20% (v/v) FCS, 19% (v/v) dimethyl sulfoxide (DMSO) (Sigma-Aldrich)) at concentration of 10^6 cells per ml, transferred to cryogenic vials and placed at -70^{0}C. After overnight incubation, cells were replaced in liquid nitrogen for storage. To bring the frozen cells again into culture, they were thawed in a water bath at 37^{0}C and resuspended in normal medium. After 16 h the culture medium was replaced to remove residual DMSO, and the cells were grown as usual.

3.2.4.4. Transient transfection of mammalian cells

LLC-PK$_1$ cells were transfected with GFP-pRS1, or β-gal-GFP with inserted fragments of pRS1 using FuGENE 6 Transfection Reagent (Roche Diagnostics, Mannheim, Germany) following the instructions of the manufacturer. For the protein purification experiments, cells were transiently transfected with GFP-CK2-NS-PKC-PKC-β-gal using polyethylenimine (Sigma-Aldrich) (Ehrhardt et al., 2006). With this method the transfection efficiency was considerably higher that enabled purification of higher amounts of the protein. For 10 cm Petri dish, a transfection mix of 18 μg of DNA, 1.2 ml of DMEM medium without any additives, and 42 μl of 1 mg/ml PEI was prepared. After 30 min, the transfection mix was added to the cells and 5 h later the cell medium was replaced by normal cultivation medium. Transfection efficiency was verified the day after according to GFP fluorescence.

3.2.4.5. Generation of stable cell lines

Stable cell lines were generated by clonal selection in the presence of 1 mg/ml gentamycin. For stable transfection subconfluent HEK cells were transfected using lipofectin reagent (GibcoBRL, Karlsruhe, Germany) following the instructions of the manufacturer. 48 h after transfection, drug selection was initiated by addition of 0.2 mg/ml gentamycin to the medium. Concentration of the antibiotic was increased by 0.2 mg/ml every 3-4 days up to 1 mg/ml. Selection medium was replaced on a daily basis. Six days after drug selection, extensive cell death was observed. Two-three weeks after the beginning of drug selection, surviving cells started forming colonies. Approximately 2-3 weeks later, single colonies were picked with a sterile pipette tips, broken up onto a single cell suspension by repetitive pipetting, and expanded. GFP-expressing clones were analyzed by fluorescent microscopy, and clones showing an appropriate GFP expression level were used for further experiments.

3.2.4.6. Inhibitor treatment of cells

In some experiments cells were treated with agents that are listed in Table 5.

Table 5. Agents and inhibitors used in this study. The final concentration and the time of incubation are indicated.

Inhibitor/ Activator	Description	Final concentration	Time of incubation
Leptomycin B	Nuclear export inhibitor	10 nM	2.5 h
Phorbol 12-myristate 13-acetate	PKC activator	0.1 µM	30 min
MG-132	Proteasome and calpain inhibitor	10 µM	20 h
MG-262	Proteasome inhibitor	0.2 nM	20 h
calpeptin	Calpain inhibitor	50 µM	20 h
A23187	Calcium ionophore	1 µM	30 min
W-13	Calmodulin inhibitor	10 µg/ml	30 min
Mimosine	Arrests cells in M phase	1 mM	16 h
Nocodazole	Arrests cell in late G1 phase	1 µM	12 h

3.2.5. Analysis of gene expression in mouse embryonic fibroblasts (MEFs)

3.2.5.1. Isolation and cultivation of MEFs

Heterozygous $RSI^{+/-}$ mice (Osswald et al., 2005) were crossed to obtain wild-type and null $RSI^{-/-}$ embryos. Early the following morning the female animals were checked for a vaginal plug which is an indicator that coitus occured. The plug is made of coagulated secretions from the coagulating and vesicular glands of the male. It generally fills the female's vagina and persists for 8-24 hours after breeding. To see the plug, the females were lifted by the base of their tails and their vaginal openings were examined for a whitish mass. Female mice having the plugs were separated. A pregnant female mouse was sacrificed at day 13-14 p.c. by cervical dislocation. The uterine horns were dissected out, and the embryos were subsequently released. Placenta, surrounding membranes and the visceral tissue were dissected. The brain was cut off and used for PCR analysis. The rest fetal tissue was rinsed in PBS (Sigma-Aldrich), minced, and treated with trypsin-EDTA (Hanks' Balanced Salt Solution containing phenol red, 0.25% porcine trypsin, and 1mM EDTA, Sigma-Aldrich) for 10 min at 37^0C. Trypsinisation was stopped by addition of an equal volume of MEF medium (DMEM supplemented with 4.5 g/l glucose (Sigma-Aldrich) containing 10% (v/v) FCS (Sigma-Aldrich), 1% L-glutamine (PAA) and 1% penicillin/streptomycin (PAA)). Thereafter, the cells were spun down and plated in a 100 cm^2 dish. These cells were defined as passage # 0. After 24-48 h, the cells became confluent. These cells were frozen or further cultivated. For continuous culturing, MEF cultures were split 1 : 5. MEFs were grown at 37^{0}C in a humidified 5% CO_2 atmosphere in the MEF medium.

3.2.5.2. Cryoculture

Detached and sedimented MEFs were resuspended in cryomedium (DMEM containing 4.5 g/l glucose, 20% (v/v) FCS, 19% (v/v) dimethyl sulfoxide (DMSO) (Sigma-Aldrich)), transferred to cryogenic vials and placed at -70^{0}C. After overnight incubation, cells were replaced in liquid nitrogen for storage. To bring the frozen MEFs again into culture, they were thawed as rapidly as possible and resuspended in normal MEF cultivation medium. After 16 h the culture medium was replaced to remove residual DMSO, and the cells were grown as usual.

3.2.5.3. Synchronization of MEFs

MEFs were synchronized to the G_1/G_0 phases of the cell cycle by incubation with DMEM (Sigma-Aldrich) containing 0.5% (v/v) FCS (Sigma-Aldrich), 1% L-glutamine (PAA) and 1% penicillin/streptomycin (PAA) for 24-36 h at 37^{0}C in a humidified 5% CO_2 atmosphere. The MEFs were subsequently exposed to normal growth medium containing 10% FCS for 12 h to enter the S phase (Jackman and O'Connor, 2001). These cells were collected for the total RNA isolation.

3.2.5.4. Isolation of Total RNA

Isolation of the total RNA was performed with RNeasy Midi Kit (Qiagen) according to manufacturer's instructions. 1-5 x 10^6 fibroblasts were used for each preparation. The concentration of RNA was determined (3,2,1,15,) and adjusted to 1 µg/µl, and the samples were stored at -80^{0}C.

3.2.5.5. Gene expression microarray analysis

The analysis of the quality of RNA, microarray analysis and normalization of the raw data were performed by Dr. Susanne Kneitz (Interdisciplinary Centre for Clinical Research, Institute of immune biology). The quality of RNA was analyzed using capillary electrophoresis (Agilent 2100

Bioanalyzer, Agilent Technologies, Palo Alto, CA). Purified total RNA samples were used for gene expression microarray experiments with GeneChip Mouse Gene 1.0 ST Array Kit (Affymetrix, Munich, Germany). Embryos with $RSI^{-/-}$ and $RSI^{+/+}$ genotype were derived from the same animal, and three wild-type – knock-out pairs from three different animals were analysed. For the evaluation of normalized data, Gene Ontology terms were used (Ashburner et al., 2000).

3.2.6. Fluorescence analysis and measurements of nuclear localization

Transfected cells grown on glass slides or polyester membranes (Corning, Duesseldorf, Germany) were washed three times with PBS buffer (137 mM NaCl, 2.7 mM KCl, 1.5 mM KH_2PO_4, 8 mM Na_2HPO_4) (Sigma-Aldrich) and fixed with 4% (w/v) paraformaldehyde in PBS for 15 at RT. Nuclei were stained with 10 µg/ml 4,6-diamindino-2-phenylindole (DAPI, Molecular Probes, Eugene, USA) which forms fluorescent complexes with nuclear double-stranded DNA (Kubista M. et al., 1987). After two washing steps with PBS buffer, cells were embedded in Fluorescent-Mounting Medium DAKO (Diagnostika GmbH, Hamburg, Germany) and placed onto the glass slides.

Cells were examined by the conventional fluorescence microscopy using an Axioplot 2 microscope (Carl Zeiss) with standard filter set. For detection of GFP fluorescence, laser excitation was performed at a wavelength 488 nm and emission was registered at 505 nm using a long pass filter. For detection of DAPI-stained nuclei, laser excitation was performed at a wavelength 345 nm and emission was registered at 458 nm. In each experiment, 100-300 transfected cells were evaluated for GFP fluorescence in the nuclei in the fluorescence microscope. Use of the nuclear marker DAPI insured proper identification of the cytoplasmic and nuclear compartments. Cells were scored according to the subcellular localization of GFP fusions as predominantly nuclear (with strong GFP fluorescence within nuclei) or cytoplasmic (weak or no nuclear GFP fluorescence). Weak nuclear fluorescence in the light microscope is due to GFP in the cytosol compartment above or below the nuclei as shown by confocal laser scanning microscopy and counterstaining with DAPI (Leyerer, 2007). Nuclear location was determined as a percentage of transfected cells containing a fluorescent protein in the nucleus.

3.2.7. Calculation and Statistics

Each individual experiment was repeated independently three to six times. Mean values ± SD were calculated. One-way ANOVA test with post hoc Tukey comparison was used to test for significance of differences between mean values from three or more groups. Significance of the differences between two mean values was calculated using Student´s unpaired t-test.

4. Results

4.1. Analysis of nuclear location of pRS1 and its fragments: experimental design

In the present study we investigated mechanisms underlying nuclear transport of pRS1 in subconfluent and confluent LLC-PK$_1$ cells. LLC-PK$_1$ cells derived from porcine epithelia serve as a popular model system for investigation of epithelial development (Hull *et al.*, 1976;Mullin *et al.*, 1980;Amsler, 1994). Subconfluent LLC-PK$_1$ cells have properties of non-differentiated cells whereas confluent LLC-PK$_1$ cells polarize and form junctions resembling differentiated epithelia.

We analysed nuclear location of pRS1 and its fragments in subconfluent LLC-PK$_1$ cells grown to 50-70% confluence (exponential growth phase (Korn *et al.*, 2001)) and in LLC-PK$_1$ cells grown four days after confluence. These two conditions will be referred throughout the text as subconfluence and confluence, respectively. In our experiments, fusion of pRS1 to the C-terminus of GFP allowed direct tracing of the protein in the cells by fluorescent microscopy. GFP tagging of pRS1 does not influence the nuclear localization of pRS1 (Kroiss *et al.*, 2006) and thus can be used for the analysis. Since small proteins ($\leq$ 50 kDa) can passively diffuse through the nuclear pore complex even in the absence of NLS (Gorlich, 1998;Rosorius *et al.*, 1999), small fragments of pRS1 were inserted between GFP and β-galactosidase which increases molecular weight of fusion proteins, thus preventing their passive diffusion.

Nuclear location was determined as a percentage of transfected cells containing a fluorescent protein in the nucleus (Yagita *et al.*, 2002;Franca-Koh *et al.*, 2002;Kobayashi *et al.*, 2003;Karlsson *et al.*, 2004). In all experiments on nuclear location, cells were transiently transfected with an indicated construct. Transient expression allowed short term analysis of different protein fragments and simplified the detection due to overexpression of corresponding GFP tagged proteins.

4.2. Dynamic redistribution of pRS1 during the cell cycle

Localization of pRS1 in LLC-PK$_1$ cells depends on confluence (Korn *et al.*, 2001;Kroiss *et al.*, 2006). Confirming these data, we observed GFP-pRS1 in 86 ± 3% (n = 6) of nuclei in subconfluent cells and in 18 ± 6% (n = 4) of nuclei in confluent cells.

Transition of LLC-PK$_1$ cells from subconfluent to confluent state is accompanied by changes in the cell density and cell cycle. Subconfluent cells actively divide and progress through G_1, S, G_2, and M phases whereas confluent cells enter the non-proliferative state (G_0 phase of the cell cycle) and develop extensive cell-cell contacts including tight junctions (Amsler, 1994).

Therefore, dependence of pRS1 localization on confluence might be due to two different reasons. First, pRS1 redistribution might be regulated by cell-cell contacts. Second, pRS1 localization might be regulated by the cell cycle. To determine whether nuclear location of pRS1 is dependent on cell proliferation/cell cycle or cell density/cell-cell contacts, we synchronized subconfluent LLC-PK$_1$ cells. Cell synchronization was required to enrich population of cells being in a specific cell cycle phase since cell populations divide asynchronously. The cells expressing GFP-pRS1 were treated with mimosine, nocodazole, or cultivated under serum withdrawal conditions, and nuclear location of GFP-pRS1 was analysed (Figure 1). Mimosine is a plant amino acid that reversibly blocks cell cycle progression at the late G$_1$ phase. When mimosine is removed, cells progress through S phase and eventually enter G$_2$/M phases. Nocodazole treatment causes destabilization of microtubules, thereby inhibiting formation of the mitotic spindle which blocks proliferation and leads to arrest in M phase. Serum starvation leads to arrest of cells in G$_1$/G$_0$ phases (Jackman and O'Connor, 2001).

In the G$_2$/M-phase generated by 12 h incubation of cells with 1 µM nocodazole (Jackman and O'Connor, 2001) virtually the same nuclear location of pRS1 was obtained as in the untreated cells (81 ± 3% versus 87 ± 2%). When the cells were synchronized in the G$_1$/G$_0$ phase by serum deprivation (Jackman and O'Connor, 2001), nuclear location of GFP-pRS1 was reduced to 27 ± 7% (Figure 1). This value is similar to nuclear location obtained in confluent cells (18 ± 6%, Figure 2). Upon synchronization in the G$_1$ phase by 16 h incubation with 1 mM mimosine (Jackman and O'Connor, 2001), 33 ± 5% of transfected cells contained GFP-pRS1 in the nuclei. To allow cell cycle progression, mimosine was removed and the cells were cultivated for 3 or 6 h without mimosine thus reaching S phase (Jackman and O'Connor, 2001). With increasing cultivation times after removal of mimosine, nuclear location of GFP-pRS1 increased successively (Figure 1). In cells grown for 9 h after removal of mimosine and, thus, progressed into G$_2$/M phase (Jackman and O'Connor, 2001), further increase in nuclear location of pRS1 could be observed (nuclear location 64 ± 7%, Figure 1). The data indicate that nuclear location of pRS1 changes during progression of cells through the cell cycle reaching a maximum in the G$_2$ and/or in the beginning of the M phase and being minimal in the G$_1$- and/or G$_0$-phases. The results indicate that confluence dependent regulation of pRS1 localization is determined, at least partially, by the cell cycle. Subconfluent cells progress through all cell cycle phases whereas the nuclear location of GFP-pRS1 in subconfluent cells was identical only to that in subconfluent cells synchronized to the M phase of cell cycle. It indicates that cell-cell contacts can also play a role in the regulation of pRS1 nuclear location.

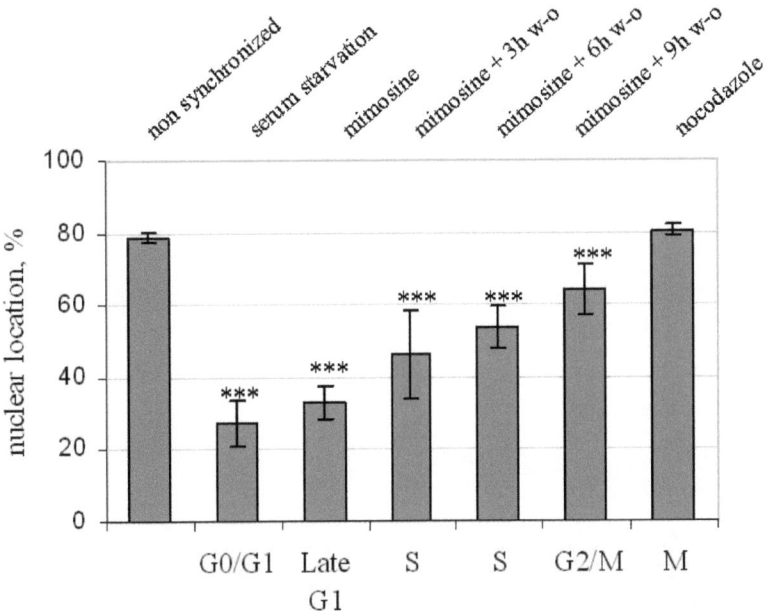

Figure 1. The nuclear location of pRS1 depends on the cell cycle. LLC-PK$_1$ cells were grown until 20-30% confluence, transfected with GFP-pRS1 and cultivated for 23 h. Subsequently cells were cultivated for 25 h under the same conditions (non synchronized), 25 h in the presence of serum depleted medium (serum starvation), 13 h with regular medium plus 12 h in the presence of 1 µM nocodazole, 9 h with regular medium plus 16 h in the presence of 1 mM mimosine (mimosine), 6 h with regular medium plus 16 h with mimosine and 3 h with regular medium (mimosine + 3 h w-o), 3 h with regular medium plus 16 h with mimosine and 6 h with regular medium (mimosine + 6 h w-o), or 16 h with mimosine and 9 h with regular medium (mimosine + 9h w-o). Fractions of transfected cells with nuclear fluorescence were determined. Mean values ± SD of four independent experiments are indicated. Synchronization conditions under which nuclear location of GFP-pRS1 was significantly different compared to non-synchronized cells are indicated by asterisks, ***$P<0.001$.

4.3. Identification and characterization of nuclear export signal in pRS1

Since the molecular weight of pRS1 (~70 kDa) surpasses the upper limit for passive diffusion through the nuclear pore (~40-50 kDa) (Gorlich, 1998;Rosorius et al., 1999), nuclear transport of pRS1 as well as redistribution of pRS1 in confluent LLC-PK$_1$ cells should occur via specific signals that facilitate its active nuclear import and possibly export. By means of mutational analysis, nuclear localization signal of pRS1 (RS1-NLS) represented by amino acids 349-369 had been previously identified (Leyerer, 2007). To identify putative nuclear export signal (NES) motif(s), a bioinformatics-based analysis employing web-based software Minimotif Miner (Balla et al., 2006) was conducted. Screening of the pRS1 protein sequence revealed the presence of a leucine-rich NES (RS1-NES), which is represented by amino acids 360-368 (^{360}LKELHELLV^{368}) (Figure 2). The sequence is consistent with the consensus sequence $\Phi X_{(2-3)}\Phi X_{(2-3)}\Phi X\Phi$ of classical nuclear export signals, where Φ is leucine or another hydrophobic amino acid, and X represents any amino acid (Wen et al., 1995;Kutay and Güttinger, 2005) The leucine-rich NESs are exported from the nucleus via interaction with the nuclear export receptor CRM1. This interaction can be blocked by leptomycin B (LMB), a potent highly specific inhibitor of the nuclear export. The inhibition mechanism involves selective alkylation of a single cysteine residue in the central conserved region of CRM1 by LMB, which impairs the binding and translocation of the target proteins (Nishi et al., 1994;Ullman et al., 1997;Kudo et al., 1999;Henderson and Eleftheriou, 2000). LMB was employed to verify whether the putative NES is functional and whether nuclear export is involved in regulation of pRS1 localization.

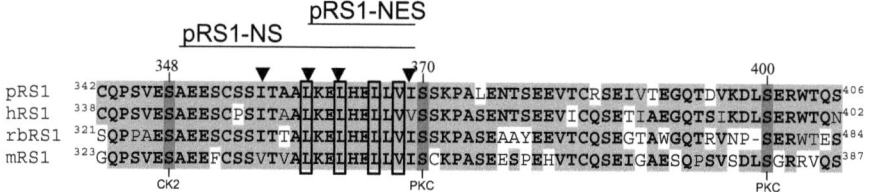

Figure 2. Alignment of the nuclear shuttling signal of pRS1 and adjacent amino acids to human RS1 (hRS1), rabbit RS1 (rbRS1) and mouse RS1 (mRS1). Light grey shadings indicate identical and similar amino acids; identical amino acids are also shown in bold face. Amino acids that represent the consensus sequence for nuclear export (NES) are boxed. Amino acids comprising the consensus motif 1-8-14 of Ca^{2+} dependent calmodulin binding are indicated by arrowheads. Regions corresponding to nuclear export signal (NES) (amino acids 360-368 of pRS1) and nuclear shuttling signal (NS) (amino acids 349-369 of pRS1) are depicted above the sequence alignment. Dark grey shadings indicate the serines 348, 370 and 400 of pRS1. Alignment of NS sequences in porcine RS1 and its orthologs was performed with Clustal X (Version 1.83) (see Methods).

LLC-PK$_1$ cells were grown until 20-30% confluence or two days after confluence, transiently transfected with GFP-pRS1, cultivated for another two days and incubated for 2.5 h with 10 nM LMB. This concentration was shown to effectively inhibit CRM1-mediated nuclear export (Sachdev and Hannink, 1998;Tyagi et al., 1998;Asscher et al., 2001). In the subconfluent cells, most of which contained GFP-pRS1 in the nucleus, treatment with LMB further increased the nuclear location of GFP-pRS1 with borderline significance (from 86 ± 2.4% to 96 ± 1.6%, P = 0.054; Figure 3). In the confluent cells, the effect of LMB was more pronounced and highly significant. 18 ± 7.2% of the confluent cells contained GFP-pRS1 in the nucleus and LMB increased nuclear location of GFP-pRS1 to 77.5 ± 1.3% (Figure 3), a value similar to the nuclear location of GFP-pRS1 in subconfluent cells. The data indicate that LMB prevents the decrease of the nuclear location of GFP-pRS1 in confluent LLC-PK$_1$ cells by inhibition of the nuclear export. Therefore, confluence may accelerate the nuclear export of pRS1 but not inhibit the nuclear import of pRS1. It suggests that the decrease of the nuclear location during confluence is due to an activation of the nuclear export mediated by CRM1.

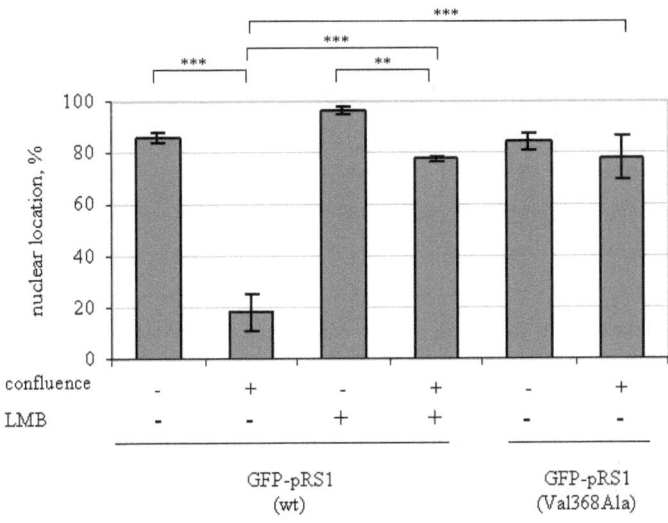

Figure 3. The decrease of nuclear location of pRS1 during confluence is abolished when the nuclear export is blocked by leptomycin B or when NES in pRS1 is inactivated. LLC-PK$_1$ were grown until 20-30% confluence or until two days after confluence and transfected with GFP-pRS1 or GFP-pRS1(Val368Ala) containing an inactivated NES. After further cultivation for two days cells were treated for 2.5 h with 10 nM LMB to inhibit CRM1 dependent nuclear export. Cells that were not incubated with LMB served as controls. Fractions of transfected cells with nuclear staining were determined. Mean values ± SD of four independent experiments are indicated. **$P<0.01$, ***$P<0.001$.

We assumed that targeting of βGal-NS-GFP into the nucleus occurs due to the specific signal mediating nuclear translocation, pRS1-NLS. However, it was formally possible that pRS1-NLS possesses a cleavage site for a protease, and cleavage with this protease yields two products: GFP fused to a fragment of the NLS and β-galactosidase fused to the rest part of the NLS. In this case, the observed nuclear GFP fluorescence would reflect the nuclear location of GFP fused to the degradation product of the NLS which can passively diffuse into the nucleus even in the absence of the nuclear import function. However, purification of the fusion protein of GFP-βGal with pRS1 fragment containing pRS1-NLS with anti-GFP antibodies yielded only the full-length protein, and no degradation products could be detected with anti-GFP antibodies on the Western blot (see subsection 4.7; Figure 7). Likewise, Westen blot analysis of the lysates of LLC-PK$_1$ cells transfected with GFP-pRS1 with anti-GFP antibodies revealed the presence of only the full length protein in subconfluent and confluent cells (see Appendix III, Figure 13). Thus, by analyzing GFP fluoresence we tracked only the full-length proteins which cannot surpass the nucleus passively. These data indicate that pRS1-NS mediates the transport of a heterologous protein into the nucleus and contains a functional NLS. Moreover, our results suggest that the differential localization of pRS1 in confluent *versus* subconfluent cells is not due to degradation of the protein but due to specific regulation of the signals governing pRS1 nuclear transport.

4.4. hRS1 interacts with nuclear import receptor importin β1

Having shown that pRS1-NES within pRS1-NS enables CRM-1-mediated export, we tried to identify the nuclear import pathway used by pRS1-NLS. The nuclear import mechanism depends on interaction between the transport receptors importins and NLSs of cargos (Pemberton and Paschal, 2005). The majority of proteins bearing nonclassical NLSs bind directly to specific subtypes of importin β family of the nuclear import receptors (Pemberton and Paschal, 2005). Therefore, we investigated whether a fragment of hRS1 containing NS (hRS1-NS-fragment, amino acids 338-402) or full-length hRS1 bind to importin β1 and/or β2 (Figure 4), the nuclear import receptors that translocate majority of targets with nonclassical NLS (Pemberton and Paschal, 2005). These experiments were performed with human RS1 in human embryonic kidney (HEK) 293 cells because the human nuclear transport machinery is well characterized and antibodies recognizing porcine importins were not available. For co-immunoprecipitation, subconfluent HEK 293 cells transfected with GFP-TEV-S-tag-(hRS1-NS-fragment) (Figure 4a) or yellow fluorescent protein (YFP)-tagged full-length hRS1 (hRS1-YFP) (Figure 4b) were employed. HEK 293 cells expressing GFP-TEV-S-tag (Figure 4a) or nontransfected HEK 293 cells (Figure 4b) served as controls. Immunoprecipitation of complexes was performed with commercial antibodies raised against GFP that also recognize YFP. The detection of the importins in bound fractions was accomplished using antibodies against importin β1 or importin β2. The reactivity of the antibodies was confirmed by the immunoblotting detection of importin β1 and importin β2 proteins in the HEK 293 lysates (data not shown). Co-precipitation of importin β1 with hRS1-NS-fragment but not with GFP transfected control cells was observed whereas importin β2 was not detected in the precipitates (Figure 4a). To demonstrate the interaction of importin β1 with full-length hRS1 protein, we precipitated hRS1-YFP from transiently transfected HEK 293 cells using commercial antibody against GFP. Consistent with the migration of hRS1 at 100 kDa in SDS-PAGE gels (Veyhl *et al.*, 2006), hRS1-YFP protein migrated at about 130 kDa (Figure 4b, left part). Similarly to GFP-TEV-S-tag-(hRS1-NS-fragment), hRS1-YFP co-precipitates contained importin β1. The data indicate that nonconventional nuclear location signal in NS of hRS1 associates with importin β1 which can mediate nuclear import of hRS1.

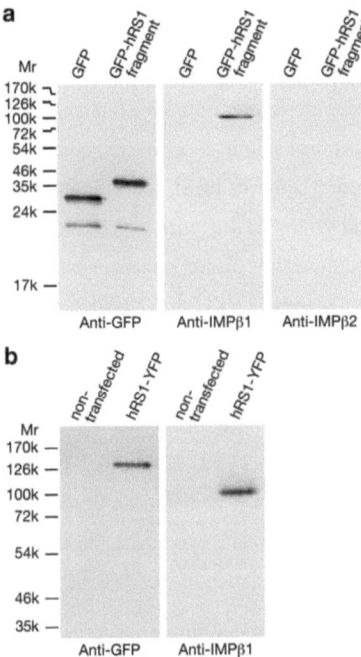

Figure 4. Interaction of the nonconventional nuclear location sequence in RS1 with importin β1. HEK 293 cells were stably transfected with GFP (GFP), GFP fused to the CK2-NS-PKC-PKC fragment of hRS1 (amino acids 338-402 of hRS1, GFP-hRS1 fragment) or YFP fused to full-length hRS1 (hRS1-YFP). Cells were lysed by sonication, centrifuged at 100 000g, and the supernatants were used for immunoprecipitation of GFP and GFP protein complexes with anti-GFP beads. Gel-separated proteins were analyzed by Western blotting using antibody against GFP (anti-GFP), importin β1 (anti-IMPβ1), or importin β2 (anti-IMPβ2). The data indicate that importin β1 is coprecipitated with the hRS1 fragment containing NLS of hRS1 and with full-length hRS1.

4.5. Identification of a minimal sequence steering confluence dependent location of pRS1

The composition of pRS1-NS containing both NLS and NES suggests that it might ensure confluence dependent nuclear migration of pRS1. However, the distribution of the βGal-GFP containing the isolated NS did not depend on confluence (Leyerer, 2007); Table 6). It suggests that an additional regulatory element(s) in pRS1, which is not present in the isolated NS, is required for the activation of NES and ensures confluence dependent nuclear transport. Since phosphorylation is one of the major mechanisms of the nuclear transport regulation (Poon and Jans, 2005), we investigated whether consensus sequences for phosphorylation surrounding pRS1-NS are required for confluence dependent nuclear location. pRS1 comprises three consensus sequences for phosphorylation in the vicinity of NS: a consensus sequence for casein kinase 2 (CK2) dependent phosphorylation of serine 348 N-terminally adjacent to NS and two consensus sequences for protein kinase C (PKC) dependent phosphorylation of serine 370 (adjacent to NS) and serine 400 (Figure 2). The three consensus sequences are conserved among human, pig, rabbit and mouse RS1 (Figure 2) suggesting their potential functional significance. We hypothesized that phosphorylation of these sites might be involved in the regulation of pRS1 subcellular distribution during confluence. To check this hypothesis and to find a minimal sequence required for the confluence dependent nuclear targeting of pRS1, we generated βGal-GFP fusion proteins containing NS and the N-terminal CK2 site (CK2-NS), NS and the first C-terminal PKC site (NS-PKC), NS and two PKC sites (NS-PKC-PKC), NS and the CK2-site and the first C-terminal PKC site (CK2-NS-PKC), and NS and all three phosphorylation sites (CK2-NS-PKC-PKC), and measured their nuclear location in the subconfluent and confluent cells (Table 6). Confluence dependent decrease of nuclear location was observed with NS-PKC, CK2-NS-PKC, NS-PKC-PKC and CK2-NS-PKC-PKC. The data indicate that NS-PKC is the minimal sequence that mediates confluence dependent nuclear location and imparts this property to the full-length protein. Moreover, the data suggest that phosphorylation of serine 370 is critically involved in the regulation of the pRS1 nuclear location during confluence.

Table 6. A fragment containing the nuclear shuttling sequence of pRS1 extended C-terminally to five amino acids is able to mediate confluence dependent nuclear location of βGal-GFP. Nuclear location of βGal-GFP containing the indicated fragments of pRS1 was measured in subconfluent and confluent LLC-PK$_1$ cells. Cells were grown until 20-30% confluence or two days after confluence and transfected with βGal-GFP containing the indicated fragments. 48 h after transfection cells were evaluated for nuclear GFP fluorescence. Mean values ± SD of three to five independent experiments are shown. *P <0.05, **P<0.01, ***P<0.001: differences to βGal-NS-GFP (NS) in confluent cells; •P<0.05: difference between NS-PCK and CK2-NS-PKC, ••• P<0.001: difference between S370 A and S370E; ˙P<0.05, ˙˙P<0.01, ˙˙˙ P<0.001: differences between subconfluent and confluent cells. The data indicate that NS-PKC is sufficient to mediate confluence dependent nuclear targeting.
a – data from Leyerer, 2007

Name	Fragments (aa)	Nuclear location	
		Subconfluent	Confluent
NS [a]	349-369	56 ± 9.0	69 ± 16
CK2-NS	342-369	44 ± 8.5	42 ± 4.5*
NS-PKC	369-374	66 ± 6.7	36 ± 7.1**˙˙
CK2-NS-PKC	342-374	62 ± 5.7	16 ± 6.3***˙˙˙•
NS-PKC-PKC	349-406	59 ± 9.5	32 ± 4.4**˙
CK2-NS-PKC-PKC [a]	342-406	67 ± 7.2	20 ± 2.4***˙˙˙
S370A [a]	342-406	50 ± 5.7	61 ± 3.8
S370E [a]	342-406	7.6 ± 2.3	21 ± 1.8***•••

4.6. Investigation of the role of phosphorylation of serine 370

The role of phosphorylation of serine 370 in nuclear targeting of pRS1 was examined using mutants of the full-length pRS1 and CK2-NS-PKC-PKC in which serine 370 was substituted by alanine or glutamate. Alanine (Ser370Ala) substitution was constructed to abolish phosphorylation at serine 370, whereas glutamate (Ser370Glu) mutants were designed to mimic a constitutively phosphorylated serine 370 (Thorsness and Koshland, Jr., 1987;Zhao *et al.*, 1994). It has been shown previously that substitution of serine 370 to alanine or glutamate leads to the loss of confluence-sensitivity of nuclear location; at that, the effects of the mutations were opposite. Substitution of serine 370 by alanine led to the nuclear accumulation of the proteins in both subconfluent and confluent LLC-PK$_1$ cells; in contrast, the Ser370Glu mutants showed low nuclear location in subconfluent and confluent cells (Leyerer, 2007). The data suggested that phosphorylation of serine 370 decreases nuclear location and might be responsible for the redistribution of pRS1 between the nucleus and the cytoplasm during confluence. Since confluence-dependent redistribution of pRS1 is dependent on the regulation of the nuclear export (see subsection 4.3), phosphorylation might be the mechanism underlying the activation of pRS1 nuclear export. Thus, we analyzed the effect of LMB on the nuclear location of the mutated (Ser370Glu) full-length pRS1. LMB increased nuclear location of the mutant in the subconfluent and confluent cells (confluent cells without LMB: 17 ± 5.9%, with LMB: 75 ± 14.3%, n = 4 each, P<0.05; subconfluent cells without LMB 16.5 ± 13.5%, with LMB 86 ± 2.8%, n = 3 each, P<0.001) (Figure 5). Therefore, phosphorylation of serine 370 promotes nuclear export of the full-length pRS1.

The effect of LMB treatment was different in case of (Ser370Glu) mutant of the CK2-NS-PKC-PKC fragment. The fragment showed no statistically significant increase in nuclear location after LMB treatment (confluent cells without LMB 21 ± 1.8%, with LMB 21 ± 1.4%, n = 3 each (Figure 5); subconfluent cells without LMB 7 ± 2.1%, with LMB to 16.5 ± 13.5%, n = 3 each) indicating that phosphorylation of serine 370 in fragment CK2-NS-PKC-PKC disturbs the nuclear import. However, since nonphosphorylated mutant of CK2-NS-PKC-PKC (Ser370Ala) was located in the nucleus independently of confluence (Leyerer, 2007; Table 6), phosphorylation of serine 370 is supposed to be also essential for the nuclear export of this fragment.

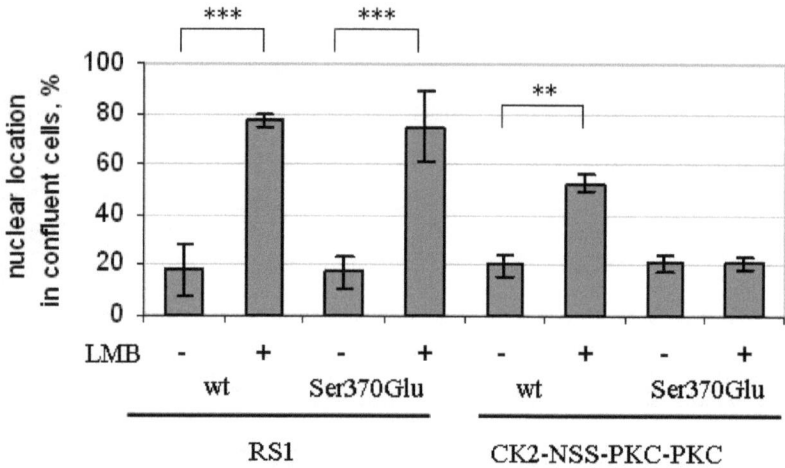

Figure 5. Influence of serine 370 phosphorylation on the nuclear location of the full-length RS1 and CK2-NS-PKC-PKC in confluent LLC-PK$_1$ cells. LLC-PK$_1$ cells were grown until two days after confluence and transfected with GFP-pRS1, βGal-CK2-NS-PKC-PKC-GFP, or a corresponding protein containing Ser370Glu mutation. After further cultivation for two days cells were treated for 2.5 h with 10 nM LMB to inhibit CRM1 dependent nuclear export. Cells that were not incubated with LMB served as controls. Fractions of transfected cells with nuclear staining were determined. Mean values ± SD of at least three independent experiments are indicated. **P<0.01, ***P<0.001.

Importantly, unlike the constitutively phosphorylated fragment, the nuclear import of the wild-type CK2-NS-PKC-PKC was not blocked, and the nuclear location of the wild-type fragment in the confluent cells was increased after LMB treatment (without LMB 20 ± 2.4%, with LMB 53 ± 3.6%, n = 3 each, P<0.001) (Figure 5). The similarity of the LMB effect on the nuclear location of the full-length pRS1 and CK2-NS-PKC-PKC indicates that nuclear location of the fragment is regulated during confluence by the nuclear export as that of the full-length protein. The different effects of LMB on the nuclear location of the mutated (Ser370Glu) pRS1 fragment versus mutated (Ser370Glu) full-length pRS1 may have different reasons. First, the folding of the mutated pRS1-NS-PKC domain may slightly differ between the isolated fragment and the full-length pRS1 protein. Another reason could be that an additional domain of pRS1 modifies structure and function of the mutated pRS1-NS-PKC domain. The similar effects of LMB on the location of GFP-pRS1 and GFP-pRS1(Ser370Glu) mutant support the interpretation that phosphorylation of serine 370 increases nuclear export whereas it does not block nuclear import of the full-length pRS1. The exchange of serine 370 to glutamate in fragment CK2-NS-PKC-PKC may not only stimulate nuclear export like in the full-length pRS1 but - different to the full-length pRS1 - also impair the nuclear import.

The increase in the nuclear location of the wild-type fragment but not of the (Ser370Glu) mutant fragment after LMB treatment can be explained on the basis of the following observations. Constitutively phosphorylated fragment is located predominantly in the cytoplasm, whereas the wild-type fragment can be phosphorylated in both the cytoplasm and the nucleus. It is, therefore, possible that CK2-NS-PKC-PKC is first imported into the nucleus and then becomes phosphorylated in the nucleus. It might explain why nuclear import of (Ser370Glu) mutant but not of the wild-type fragment was blocked. The (Ser370Glu) mutant of the full-length RS1 might either be phosphorylated in both the cytoplasm and the nucleus, or might bear additional regulatory elements which aid in nuclear translocation of RS1.

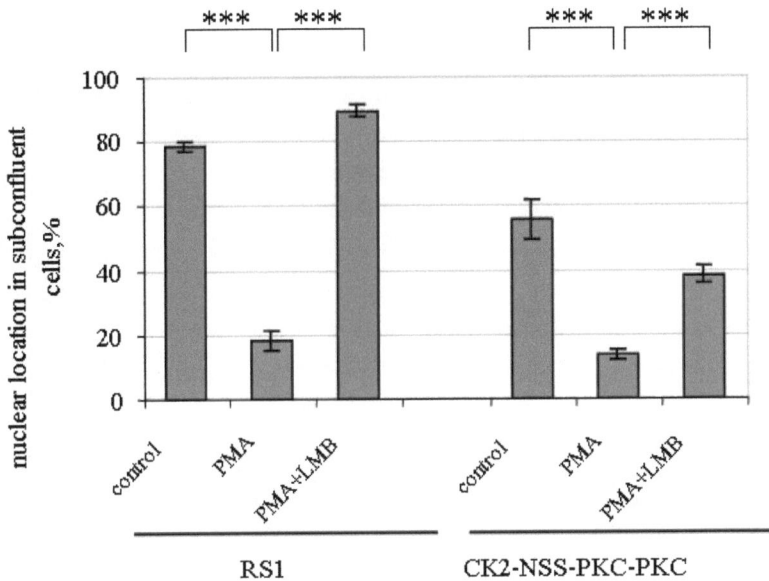

Figure 6. PKC activation leads to a decrease of the nuclear location of the full-length RS1 and of CK2-NS-PKC-PKC which can be blocked by the inhibition of nuclear export. LLC-PK$_1$ were grown until 20-30% confluence and transfected with GFP-pRS1 or CK2-NS-PKC-PKC. After further cultivation for two days cells were treated with 0.1 µM PMA or 0.1 µM PMA together with 10 nM LMB for 2.5 hours. Non-treated cells served as control. Fractions of transfected cells with nuclear staining were determined. Mean values ± SD of four independent experiments are indicated. ***P<0.001.

To prove the role of PKC in the regulation of pRS1 subcellular localization, we investigated whether application of an activator of PKC, phorbol 12-myristate 13-acetate (PMA), influences the nuclear location of GFP-pRS1 and β-Gal-(CK2-NS-PKC-PKC)-GFP in subconfluent LLC-PK$_1$ cells. Similar results were obtained with both proteins. When cells were incubated for 30 minutes with 0.1 µM PMA, the nuclear location of the corresponding proteins was significantly decreased (GFP-pRS1 from 79 ± 1.6% to 26.5 ±1.5%, β-Gal-(CK2-NS-PKC-PKC)-GFP from 67 ± 7% to 14 ± 1.5%, Figure 6). Significant shift towards a more cytoplasmic localization of pRS1 upon PKC activation was predicted from the data obtained with serine mutants and indicates that localization of pRS1 is controlled by PKC. To distinguish between the effects of PKC on nuclear import and nuclear export of pRS1, subconfluent cells were treated with 0.1 µM PMA or 0.1 µM PMA together with 10 nM LMB for 2.5 hours. Similarly to PMA treatment for 30 min, the PMA treatment for 2.5 hours led to a strong decrease of nuclear location of the proteins (GFP-pRS1 from 79 ± 1.6% to 19 ± 3%, β-Gal-(CK2-NS-PKC-PKC)-GFP from 67 ± 7% to 20 ± 12%). However, when LMB was applied to block nuclear export, the effect of PMA on the nuclear location of the full-length pRS1 and CK2-PKC-PKC was abolished (Figure 6) suggesting that PMA activates nuclear export by phosphorylation of serine 370.

Interestingly, no significant effect had been observed when subconfluent cells were incubated for 1 h with 5 µM PKC stimulator DOG (Leyerer, 2007). The observation that nuclear location depended on PMA but not on DOG implies that phosphorylation of serine 370 of pRS1 during confluence is mediated via a DOG insensitive PKC subtype that can be activated by PMA. It might occur in different situations. First, phorbol esters and DOG bind to two different sites of PKCs and may induce different states of activation (Slater *et al.*, 1994), and phosphorylation of serine 370 may require a PMA-specific activation of PKC. Second, PMA activates nuclear PKC subtypes (Thomas *et al.*, 1988;Leach *et al.*, 1989;Eldar *et al.*, 1992) which might be responsible for the phosphorylation of serine 370 exclusively in the nucleus.

4.7. Studies on the phosphorylation state of serine 370 of pRS1 in subconfluent and confluent LLC-PK$_1$ cells using mass spectrometry

Although the functional role of phosphorylation of serine 370 may be deduced from the experiments employing Ser370 mutants and PKC activation, the demonstration of differential phosphorylation of serine 370 in the subconfluent and confluent cells was essential for a final proof. To study the phosphorylation state of pRS1 *in vivo*, mass-spectrometry was applied. GFP-CK2-NS-PKC-PKC-β-Gal was purified from transiently transfected subconfluent or confluent LLC-PK$_1$ cells using magnetic beads coupled to anti-GFP antibodies and subjected to the mass-spectrometry analysis (LC-MRM-MS/MS) that was performed by Prof. Dr. Albert Sickmann (Rudolf-Virchow-Center, DFG Research Centre for Experimental Biomedicine, University of Wuerzburg). To prevent dephosphorylation or phosphorylation of serine 370 during lysis or purification, all purification steps were performed on ice and the solutions were supplemented with phosphatase and kinase inhibitors. The efficiency of purification was controlled by the silver staining of polyacrylamide gels and immunoblotting with anti-GFP antibodies (Figure 7). The absence of the GFP fusion protein in the flow-through and the enrichment of the elution fraction with the fusion protein (see representative silver-stained gel and Western blot in Figure 7) indicated highly efficient purification.

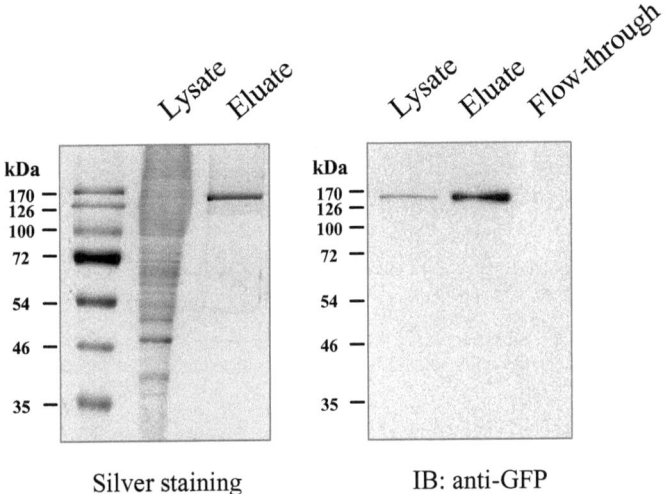

Figure 7. Purification of GFP-CK2-NS-PKC-PKC-βGal using magnetic beads coupled to antibodies recognizing GFP. The fusion protein was purified from transiently transfected subconfluent LLC-PK$_1$ cells and purified with anti-GFP antibodies. Lysate, flow-through and elution fraction were analysed by silver staining of SDS-polyacrylamide gels (*left*) and immunoblotting with anti-GFP antibodies (*right*).

The proteins GFP-CK2-NS-PKC-PKC-β-Gal purified from subconfluent or confluent LLC-PK$_1$ cells were subjected to the mass spectrometric analysis. Since mass spectrometric analysis can be applied only to peptides with molecular weight less than 3 kDa, a prior proteolytic cleavage with trypsin was performed. By mass spectrometric analysis of the trypsin cleavage products, we were unable to detect any peptide containing serine 370 – neither in phosphorylated nor in nonphosphorylated form. This could be due to a limitation in a dynamic range of mass spectrometry when a peptide might not be seen if it is present in a very low abundance relative to another peptide in the same fraction (Areces *et al.*, 2004). Other reason might be low ionization efficiency which can interfere with the detection of peptides (Areces *et al.*, 2004). Finally, the tryptic cleavage could result in generation of the peptides with molecular mass more than 3 kDa which cannot be detected by the mass spectrometry. Indeed, in our experiments the protein sequence coverage did not exceed

39% and averaged 25%. Inspection of the GFP-CK2-NS-PKC-PKC-β-Gal protein sequence revealed that two potential trypsin cleavage sites at positively charged amino acids located in the vicinity of serine 370 are followed by prolines (Figure 10) that can block tryptic cleavage (Wilkinson JM., 1986;Bier ME, 2002;Barret *et al.*, 2004). In this case, cleavage with trypsin yields too large peptides which cannot by detected by mass spectrometry. Thus, the alternative proteases including Glu-C, chymotrypsin, and V6 were employed. However, after application of these agents peptides containing serine 370 could not be detected by mass spectrometry as in the case of trypsin cleavage.

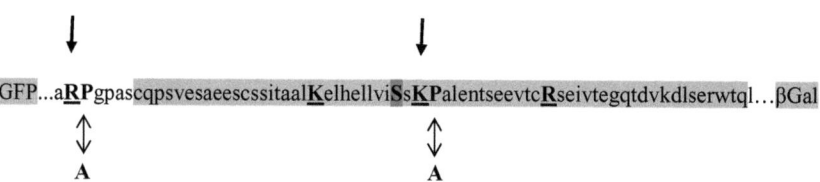

Figure 8. Alternative strategy to detect peptides containing serine 370. Prolines shown in bold were substituted with alanines, thus allowing cleavage by trypsin at two additional sites indicated by arrows. All potential trypsin cleavage sites in vicinity of serine 370 are underlined. The introduced mutations should increase the possibility that peptides containing the serine of interest (marked with dark grey) will be detected. GFP, CK2-NS-PKC-PKC (amino acids 342-402) and β-galactosidase are marked with light grey.

Since application of several cleavage agents did not produce peptides that can be detected by mass spectrometry, we developed an alternative strategy to detect peptides containing serine 370 (Figure 8). As discussed above, serine 370 is surrounded by two potential trypsin cleavage sites in GFP-CK2-NS-PKC-PKC-β-Gal, but the cleavage can be disturbed due to the proline residues following the positively charged amino acids (Wilkinson JM., 1986;Bier ME, 2002;Barret *et al.*, 2004). Therefore, proline residues at positions 247 and 283 of GFP-CK2-NS-PKC-PKC-βGal were substituted with alanines, thus allowing trypsin cleavage at two additional sites (arginine 246 in the spacer region and lysine 282 corresponding to position 361 of pRS1) surrounding the serine of interest (Figure 8). The resulting protein, GFP-CK2-NS-PKC-PKC(+2Tryp)-βGal, was expressed in subconfluent or confluent LLC-PK$_1$ cells, purified using magnetic beads coupled to anti-GFP antibodies, and subjected to nano LC-MS/MS analysis in cooperation with Dr. Yvonne Reinders (Proteomics Group, Institute of Functional Genomics, University of Regensburg). The applied strategy allowed the detection of both phospho- and non-phosphopeptides containing serine 370. These peptides had the sequence ELHELLVISSK indicating that the trypsin cleavage occurred at the natural site after Lys 361 and at the introduced site after Lys372. Integration of extracted ion chromatograms of the phosphorylated and non-phosphorylated peptides revealed that phosphorylated peptide signal was increased by approximately 60 % while the signal of the non-phosphorylated peptide was diminished by 33% in the sample derived from the confluent cells in comparison with the sample obtained from the subconfluent cells. Therefore, serine 370 is phosphorylated to a higher degree in the confluent cells in comparison with the subconfluent cells.

It is important to mention that only the ratios of abundancies of the peptide (either non-phosphorylated or phosphorylated) from subconfluent and confluent cells samples could be derived from the mass spectrometry data. The absolute quantification of the extent of serine 370 phosphorylation (e.g, ratio of phosphorylated and non-phosphorylated peptides) in proteins purified from subconfluent or confluent LLC-PK$_1$ cells was impossible since the spectra of phospho-and non-phosphopeptides generally do not accurately reflect the ratio between the phospho- and non-phosphopeptides in a sample (Craig *et al.*, 1991). The main reason is the low ionization efficiency of phosphopeptides compared with their non-phosphorylated analogues which leads to higher yields of the nonphosphorylated peptide ions compared with the phosphorylated peptide (Craig *et al.*, 1991).In parallel with mass spectrometry analysis, we aimed to generate the antibodies which recognize serine 370 specifically in phosphorylated state. The generation and characterization of these antibodies is described in Appendix IV.

4.8. Investigation of the role of calmodulin in the regulation of nuclear location of pRS1

RS1-NS contains a functional consensus motif for the calcium-dependent calmodulin binding. This motif belongs to the class of 1-8-14 motives (Rhoads and Friedberg, 1997). It overlaps with RS1-NES (Figure 2) and might overlap with determinants of RS1 nuclear import suggesting that calmodulin might be involved in regulation of RS1 nuclear transport. The nucleus has a complete Ca^{2+} signalling system and Ca^{2+} signals can generate considerable movement of calmodulin into the nucleoplasm, where it may have an important function in the control of gene expression (Petersen *et al.*, 1999). For example, it has recently been shown that during the initial phase of cholecystokinin-evoked global cytosolic Ca^{2+} oscillations, each Ca^{2+} rise causes an increment in the nucleoplasmic calmodulin concentration, which finally reaches a steady enhanced level (Craske *et al.*, 1999). To test whether calmodulin participates in confluence dependent regulation of RS1 nuclear localization, we investigated the effect of the calmodulin inhibitor W-13 on the nuclear location of GFP-pRS1 and NS-PKC in subconfluent or confluent LLC-PK$_1$ cells. This inhibitor belongs to the class of naphtalenesulfonamide derivatives which bind to calmodulin and inhibit Ca^{2+}/calmodulin regulated enzyme activities (Hidaka *et al.*, 1981). Inhibition of calmodulin binding did not change the nuclear location of RS1 in both subconfluent and confluent cells (GFP-pRS1, subconfluent cells: control, 98 ± 1.2%, with inhibitor, 99 ± 0.3%, n=3; confluent cells: control, 17.6 ± 7.3%, with inhibitor, 12.4 ± 5.1%, n=5; NS-PKC, subconfluent cells: control, 66 ± 7%, with inhibitor, 63 ± 4%, n = 3; confluent cells: control, 25 ± 8%, with inhibitor, 21 ± 8%, n = 3). It suggests that calmodulin is not involved into the confluence-dependent regulation of RS1 nuclear localization in LLC-PK$_1$ cells. The absence of the calmodulin effect on nuclear location of RS1 might have at least two different reasons. First, it is possible that calmodulin is not expressed in LLC-PK$_1$ cells or its expression level is low. Second, the intracellular free calcium concentration, estimated as 72 ± 6 nM in LLC-PK$_1$ cells (Parys et al., 1986; Blackmore et al., 2002), might be too low for calmodulin binding to RS1 since the observed half maximal calcium concentration required for calmodulin binding to RS1 in vitro has been determined as 0.48 ± 0.28 µM (experiment performed by Chakravarthi Chintalapati; Filatova et al., 2009, submitted). To assess the first possibility, we tested whether calmodulin protein is present in subconfluent and confluent LLC-PK$_1$ cells. The calmodulin was expressed in both HEK 293 cells and LLC-PK$_1$ cells (Figure 9a); therefore, it seems that low calmodulin expression cannot be the explanation for the absence of the calmodulin effect. The second possibility was tested by application of the calcium ionophore A23187 which increases the intracellular calcium concentration and, therefore, should enhance or activate calcium-dependent calmodulin binding to RS1. We analysed the nuclear location of GFP-pRS1 in subconfluent or confluent LLC-PK$_1$ cells that were treated with A23187 for 30 min. In both subconfluent and confluent cells, the elevated intracellular calcium concentration did not influence the nuclear location of RS1 (subconfluent cells, control; 98 ± 1.2%, with calcium ionophore: 99 ± 1.1%, n=3; confluent cells, control: 17.6 ± 7.3%; with calcium ionophore: 19.5 ± 2.4%, n=3) suggesting that calmodulin binding does not influence the confluence-dependent nuclear location of RS1.

To directly test whether calmodulin binding to the consensus binding motif within RS1-NS changes nuclear migration of RS1, we inactivated the binding by single substitution of isoleucine 356 to glycine or double substitution of isoleucines 356 and 369 to glycines in CK2-NS-PKC-PKC. The corresponding mutations have been shown to inactivate calmodulin binding to 1-8-14 calmodulin binding motives (Fischer et al., 1996; Beguin et al., 2005). The inactivation of calmodulin binding has been confirmed by the in vitro calmodulin binding assay (experiment performed by Chakravarthi Chintalapati; Filatova et al., 2009, submitted). Nuclear location of the mutants in subconfluent cells was strongly decreased for both mutants in comparison with the wild-type fragment (single substitution mutant: 18 ± 2.7%; double mutant: 9 ± 1.8%). The observed decline in the nuclear location of the mutants might have two reasons: either the NLS itself is destroyed, or the calmodulin binding is required for the nuclear location of RS1. Trying to distinguish between these two possibilities, we treated subconfluent or confluent LLC-PK$_1$ cells

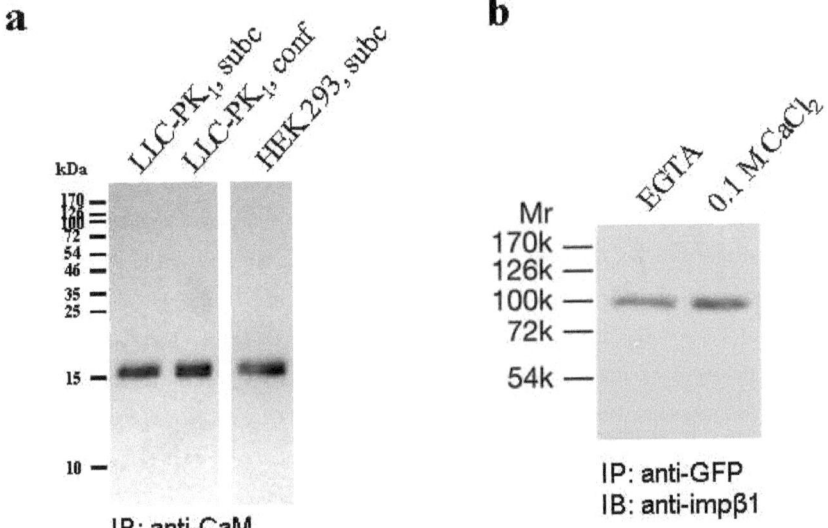

Figure 9. a. The lysates of subconfluent HEK 293 cells and subconfluent and confluent LLC-PK1 cells were analysed with an antibody against calmodulin (anti-CaM). 30 μg of protein were applied per lane. **b.** HEK 293 cells stably transfected with GFP (GFP), GFP fused to amino acids 338-402 of hRS1 (GFP-hRS1 fragment) were lysed in the presence of 2 mM EGTA or 0.1 mM Ca^{2+}, and the supernatants were incubated with anti-GFP antibody coupled agarose in the absence of Ca^{2+} (2 mM EGTA) or in the the presence of 0.1 mM Ca^{2+}. The beads were washed in the absence or presence of Ca^{2+}. GFP and GFP-protein complexes were analyzed by Western blotting using antibody against importin β1 (anti-impβ1). The data indicate that importin β1 is coprecipitated with the hRS1 fragment containing RS1-NLS in the absence and presence of Ca^{2+}.

expressing one of the mutants with LMB. Treatment with nuclear export inhibitor did not increase the nuclear location (CK2-NS-PKC-PKC (Ile356Gly): without inhibitor 18 ± 2.7%, with inhibitor 19 ± 8.2%; CK2-NS-PKC-PKC (Ile356Gly,Ile369Gly): without inhibitor 9 ± 1.8%; with inhibitor 9 ± 2.1%) implying that the mutations led to the impairment of the nuclear import. This effect can be due to the direct effect of the mutation(s) on RS1 nuclear import (e.g., when important recognition determinants of NLS are destroyed) or can be caused by disturbance of the calmodulin binding. To approach this question, we tested whether calmodulin binding prevents binding of importin β1 to RS1-NS and may thereby block nuclear import. Note that the co-precipitation of importin β1 with the human RS1 fragment CK2-NS-PKC-PKC expressed in HEK 293 cells was performed in the absence of Ca^{2+} where endogenous calmodulin does not bind to NS (see subsection 4.4). We tested whether increased calcium concentration which should enhance calmodulin binding prevents binding of importin β1 to NS and may thereby block nuclear import under these conditions.

Analysis of the precipitates of GFP-hRS1 fragment (GFP-TEV-S-Tag-(human CK2-NS-PKC-PKC) in the absence and presence of 0.1 mM Ca^{2+} showed that importin β1 was co-precipitated under both conditions, and that the amount of precipitated importin β1 was 1.97 ± 0.5 times higher (n=3, $P<0.05$) in presence of 0.1 mM Ca^{2+} compared to the calcium-free conditions (Figure 9b). The data suggest that binding of calmodulin to NS modulates but does not prevent binding of importin β1. Since the inactivation of calmodulin binding was not the reason for the decrease in nuclear location of RS1 in subconfluent cells, Ile356 seems to be important for RS1 transition into the nucleus along with the amino acid Leu366. Our data suggest that calmodulin is not involved in the confluence dependent regulation of RS1 localization. At the same time, our results do not exclude the involvement of calmodulin in the regulation of RS1 nuclear transport. The increase of the intracellular calcium concentration, for example, during confluence (Nigam *et al.*, 1992) or during the cell cycle (Bootman and Berridge, 1996;Gerasimenko *et al.*, 1996), might lead to the activation of calmodulin binding to RS1 and subsequent redistribution of the protein. This assumption, however, needs experimental support and calls for further investigation.

5. Discussion

LLC-PK$_1$, a cell line derived from porcine proximal tubule cells (Hull et al., 1976), has been widely used as a model system for studying functions of polarized epithelia and, in particular, regulation of glucose transport in epithelial cells (Mullin et al., 1980;Amsler, 1994). This cell line has been described to undergo morphological changes during differentiation and maturation from subconfluent culture to a confluent epithelial layer. During transition from logarithmically growing subconfluent undifferentiated cells to a confluent monolayer in LLC-PK$_1$ cells, a number of differentiated functional characteristics of renal proximal tubule develops including tight junctions, microvilli, vasopressin responsiveness, transepithelial salt and water transport, and brush border marker enzyme activities (Mullin et al., 1980;Amsler, 1994). Expression of apical membrane Na$^+$-coupled glucose transport activity (Amsler and Cook, 1982;Lever, 1986) and SGLT1 expression (SHIODA et al., 1994;Yet et al., 1994) is also differentiation dependent in LLC-PK$_1$ cells. SGLT1 is undetectable in subconfluent, actively dividing cultures whereas SGLT1 expression and SGLT1 mediated AMG uptake drastically increase after cell confluence together with other differentiation-specific features of kidney proximal tubule.

RS1 participates in the confluence dependent regulation of SGLT1 in LLC-PK$_1$ decreasing the expression of SGLT1 in subconfluent LLC-PK$_1$ cells via downregulation of mRNA transcripts (Korn et al., 2001). In serving this function, RS1 displays a distinct subcellular localization in subconfluent compared to confluent cells. In the subconfluent cells RS1 is present in the nucleus and the cytoplasm, whereas after confluence it is located predominantly in the cytoplasm (Kroiss et al., 2006;Leyerer, 2007). Understanding the mechanism of RS1 nuclear transport might provide a deeper insight on the mechanism of RS1-mediated regulation of SGLT1. Therefore, in the present work an attempt was undertaken to reveal a mechanism of confluence-regulated nuclear transport of RS1.

We showed that the decrease of nuclear location of RS1 observed during confluence is determined, at least partially, by the cell cycle. Therefore, RS1 might regulate SGLT1 in cell cycle dependent manner. Cell cycle dependent regulation of membrane proteins has been demonstrated previously. For example, the transcription factor c-Myb upregulates the plasma membrane Ca2+ ATPase-1 (PMCA1) during G1 / S transition (Afroze and Husain, 2001). In LLC-PK1 cells, a 32 amino acid hormone calcitonin differentially activates Na,K-ATPase and Na/H exchanger in cell cycle dependent manner (Chakraborty et al., 1991;Chakraborty et al., 1994). Likewise, cell cycle dependent nuclear location of RS1 might influence Na+-D-glucose cotransport activity during confluence of LLC-PK1 cells. However, after cell cycle arrest at G1/G0 phase in subconfluent LLC-PK1 cells induced by 24 h serum starvation or at G1 phase induced by treatment for 16 h with

1 mM mimosine leading to a decrease of nuclear location of RS1 similar to confluence, Na+-d-glucose cotransport activity was increased by only 50-75% (H. Koepsell, unpublished data). At that, Na+-d-glucose cotransport activity increases more than 100-fold during confluence (Korn et al., 2001). Therefore, the reduced amount of RS1 in the nucleus is not the main reason of the drastic upregulation of Na+-d-glucose cotransport activity during confluence of LLC-PK1 cells. Additional regulatory processes must be involved such as RS1 independent upregulation of SGLT1 transcription and/or posttranscriptional upregulation of SGLT1. The regulation of the mRNA level might be accomplished, for example, by PKA which raises SGLT1 mRNA level by a pronounced stabilization of the message (Amsler et al., 1991;Peng and Lever, 1995). The posttranscriptional upregulation may be independent of RS1 or may be due to posttranscriptional regulation of SGLT1 by RS1 (Korn et al., 2001;Veyhl et al., 2006;Kroiss et al., 2006). This additional regulation of SGLT1 during confluence may be signalled by cell-cell contact. In summary, nuclear location of RS1 is not the main regulatory mechanism for confluence dependent upregulation of SGLT1 in LLC-PK1 cells; however, changes in nuclear location of RS1 contribute to the upregulation of SGLT1 during the cell cycle and the confluence. This type of regulation may be critically involved in changes of SGLT1 expression during regeneration of enterocytes in small intestine (Freeman et al., 1992) and during regeneration of renal tubular cells after hypoxemic stress (Jiang et al., 2005). Although RS1 has been mainly described as a regulator of SGLT1 (Veyhl et al., 1993;Lambotte et al., 1996;Reinhardt et al., 1999;Veyhl et al., 2003;Osswald et al., 2005;Veyhl et al., 2006;Vernaleken et al., 2007) it must be kept in mind that RS1 regulates the expression of various plasma membrane transporters including polyspecific drug transporters of the SLC22 transporter family and sodium dependent neurotransmitter transporters (Reinhardt et al., 1999;Jiang et al., 2005). RS1 is expressed in various normal tissues, different cell types and in most tumours (Reinhardt et al., 1999;Osswald et al., 2005); see also cDNA chip expression arrays: http://www.ncbi.nlm.nih.gov/geo) and is supposed to be involved in tissue specific regulations of a group of plasma membrane transporters. We assume that the transcriptional downregulation of transporters in poorly differentiated cells is a general function of RS1 and may be important in tumour cells. Thus, the regulation of nuclear location of RS1 can aid in controlling this function.

Although a large number of proteins have been known to shuttle between the nucleus and the cytoplasm, there are only a few examples of proteins regulated by confluence. Notably, similarly to RS1, all of them are localized predominantly in the nucleus in subconfluent cells but in the cytoplasm in the confluent cells. The confluence-regulated nucleocytoplasmic shuttling proteins include several transcriptional factors such as the tumor suppressor proteins, von Hippel-Lindau (Lee et al., 1996) and adenomatous polyposis coli (APC) (Zhang et al., 2001), the aryl hydrocarbon receptor (AhR) (Ikuta et al., 2004), Smad2 (Petridou et al., 2000), and the influenza virus nucleoprotein (Bui et al., 2002) as well as proteins with an unknown nuclear function: the growth regulator p8 (Valacco et al., 2006), the junction-associated proteins ZO-1 (Gottardi et al., 1996)and ZO-2 (Islas et al., 2002), and the poorly characterized proteins AHNAK (Sussman et al., 2001) and NORPEG (Kutty et al., 2006). The confluence dependent redistribution of some of these proteins is mediated by nuclear export (Ikuta et al., 2004) or differential phosphorylation of sites near NLS or NES (Zhang et al., 2001;Sussman et al., 2001;Bui et al., 2002;Ikuta et al., 2004) as in case of RS1. However, to our knowledge, RS1 is the first described protein confluence-dependent shuttling of which is mediated by a single 26 amino acids long nuclear shuttling domain.

The mapping of a signal which controls the nuclear transport of RS1 was one of the issues addressed in this study. The nuclear shuttling signal employed by RS1 (RS1-NS) was identified which is composed of nuclear localization signal (RS1-NLS) and nuclear export signal (RS1-NES) that enable transport of RS1 in and out of the nucleus. RS1-NS is highly conserved across species showing 95% amino acid identity between pig, human, rabbit and mouse suggesting a common mechanism of nuclear import and export for pRS1 and orthologs (Figure 2).

RS1-NS is a new member of the class of nuclear shuttling signals that contain overlapping and sometimes inseparable NLS and NES (Michael, 2000;Bachmann *et al.*, 2006;Lin and Yen, 2006). It bears little sequence similarity with other characterized NS such as the M9 domain of hnRNP A1 (Siomi and Dreyfuss, 1995), KNS of hnRNP K (Michael, 2000), HNS of HuR (Fan and Steitz, 1998), or ZNS of DAZAP1 (Lin and Yen, 2006); however, RS1-NS shares the consensus of an embedded CRM1 dependent leucine-rich NES with NSs of Vpr (Sherman *et al.*, 2001), protein tyrosine kinase Syk (Zhou *et al.*, 2006), and S6 kinase 1 (Bachmann *et al.*, 2006). The existence of a subclass of NSs containing overlapping leucine-rich NES and nonconventional NLS was first proposed by Sherman et al. (2001) who have shown that leucine-rich NES of Vpr has also properties of NLS. Authors assumed that NES-containing sequences may function as NSs also in other proteins that lack a classical import signal. Our results confirm this assumption and might prompt the investigation of leucine-rich NESs of other nuclear proteins bearing unknown nuclear localization signals.

RS1-NLS shows no similarity to the classical nuclear localization signals that contain a single short stretch of basic amino acids (e.g., P*KKKRK*V (Kalderon *et al.*, 1984)) or the bipartite NLS with two basic amino acid clusters with an intervening spacer (e.g., *KR*PAATKKAGQA*KKKK* (Robbins *et al.*, 1991a)). On the contrary, RS1-NLS essentially devoid of clusters of basic residues and is represented by a putative α-helix containing hydrophobic (mainly leucine) and acidic amino acids (Figure 2). Several studies have demonstrated the involvement of the leucine-rich domains into the nuclear transport. The leucine-rich motifs mediate nuclear import of the RanGAP1 (Matunis *et al.*, 1998), sterol regulatory element binding protein

(SREBP) (Nagoshi et al., 1999;Nagoshi and Yoneda, 2001), Vpr (Sherman et al., 2001), and protein tyrosine kinase Syk (Zhou et al., 2006). Interestingly, similarly to RS1-NLS, the NLS of STEBP has a helical structure and is able to engage importin ß directly (Nagoshi et al., 1999;Lee et al., 2003a). Thus, RS1-NLS might employ the same nuclear import pathway as the NLS of STEBP. Although the full extent of RS1-NLS is not completely mapped, by means of mutational analysis we could identify amino acids Ile 356 and Leu366 as important for the nuclear import of RS1.

In the present study, the nuclear transport receptors which mediate nucleocytoplasmic shuttling of RS1 have been identified. We showed that the leucine-rich NES of RS1 employs nuclear export receptor CRM1 whereas RS1-NLS directly interacts with importin β1 that is responsible for the nuclear import of RS1. Our data suggest that the nuclear import of RS1 is not altered during confluence, and that the nucleus/cytosol distribution of RS1 is regulated during confluence by nuclear export activit.

The mechanism that regulates RS1 nuclear transport during confluence involves phosphorylation of serine 370. We assume that phosphorylation of serine 370 by PKC enhances nuclear export of RS1 after confluence basing on the following observations: (i) inactivation of phosphorylation by mutating serine 370 to alanine led to nuclear accumulation that was independent of confluence; (ii) nuclear location of GFP-RS1 and βGal-CK2-NS-PKC-PKC-GFP was strongly increased after treatment of confluent LLC-PK1 cells with nuclear export inhibitor LMB; (iii) the decline of nuclear location of GFP-RS1 and βGal-CK2-NS-PKC-PKC-GFP caused by activation of PKC could be reversed by inhibition of nuclear export. Confirming our results, mass spectrometry analysis showed that serine 370 is phosphorylated to a higher degree in confluent LLC-PK$_1$ cells in comparison with subconfluent cells. On the first view our results appear to be inconsistent with previous findings showing that total PKC activity in homogenates of subconfluent LLC-PK$_1$ cells is higher compared to confluent LLC-PK$_1$ cells (Dawson and Cook, 1987). However, the phosphorylation of pRS1 may be subtype-specific, i.e. only one specific PKC subtype may phosphorylate RS1. For example, serine 370 of pRS1 may be phosphorylated by a PKC isoform that is translocated from the cytoplasm to the nucleus in response to differentiation stimuli. Indeed, there are several reports of increased levels of nuclear PKC activity in the context of cell differentiation. All these investigations showed translocation of various PKC isoforms from the cytoplasm to the nucleus upon cell differentiation (Martelli et al., 1999;Buchner, 2000;Martelli et al., 2006). Moreover, PKC is translocated to the nucleus in response to phorbol ester treatment (Thomas et al., 1988;Leach et al., 1989;Eldar et al., 1992) that can explain the differential effects of PMA and DOG on the nuclear location of RS1 (see subsection 4.6. in Results). Although our data favour the phosphorylation of Ser370 by PKC in the nucleus, we do not have any direct evidence in which cellular compartment the phosphorylation takes place.

Taking together, the following model of regulation of confluence-dependent nuclear transport of RS1 in LLC-PK1 cells can be proposed (Figure 10). In subconfluent cells, RS1 is translocated into the nucleus via the nuclear import receptor importin β1 whereas nuclear export of RS1 is not active. RS1 accumulates in the nucleus because nonphosphorylated RS1-NS does not mediate nuclear export. After confluence, PKC phosphorylates RS1 that enhances RS1 nuclear export mediated by the nuclear export receptor CRM1. It leads to the predominantly cytoplasmic distribution of the protein in the confluent cells.

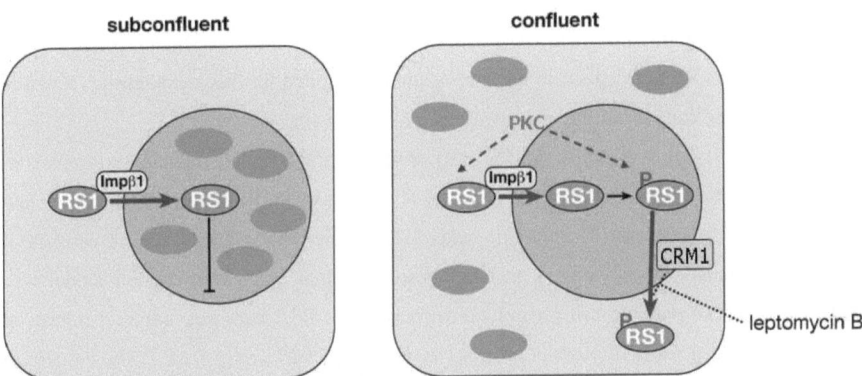

Figure 10. Schematic model for the confluence-dependent regulation of RS1 nuclear location by differential phosphorylation of serine 370. In subconfluent cells, pRS1 localises to the nucleus due to active NLS which mediates transport of RS1 into the nucleus via importin β1. In confluent cells, phosphorylation of serine 370 activates NES that leads to the expulsion of the protein from the nucleus via nuclear export receptor CRM1 and ensures predominantly cytoplasmic distribution pattern of pRS1.

The impact of Ca^+ dependent calmodulin binding to RS1-NS on the regulation of nuclear location of RS1 is not understood. We could not elucidate the role of calmodulin in confluence-dependent regulation of nuclear location of RS1. This may be due to relatively low Ca^{2+} concentrations in LLC-PK$_1$ cells (Blackmore et al., 2002). Importantly, our results do not exclude that calmodulin binds to RS1-NS under more defined physiological conditions, for example during specific cell states when the cytosolic and/or nuclear Ca^{2+} concentrations increase to higher levels. Such states may be represented only by single cells of the populations of LLC-PK$_1$ cells investigated in our study. Notably, transient large increases of Ca^{2+} concentration in the cytosol may be channelled in the nucleus during confluence of MDCK cells (Nigam et al., 1992), during antigen stimulation of tumour mast cells (Chandra et al., 1994) or during the cell cycle (Bootman and Berridge, 1996;Gerasimenko et al., 1996). Since modelling of interaction of RS1-NS with calmodulin and PKC presupposes that calmodulin and PKC cannot bind simultaneously to RS1-NS

(Filatova et al., 2009, submitted), calmodulin binding should compete with the phosphorylation of serine 370 by PKC and thereby slow down or prevent nuclear export of RS1.

The nuclear location of RS1 in confluent cells strongly increases within 2.5 hours upon inhibition of the nuclear export. It suggests that a large fraction of the total RS1 population shuttles between the nucleus and cytoplasm. What is the role of nuclear shuttling by RS1? RS1 downregulates the release of SGLT1 containing vesicles from the TGN (Veyhl *et al.*, 2006;Kroiss *et al.*, 2006) as well as the transcription of SGLT1 (Korn *et al.*, 2001). Since RS1 mediates a dual regulation of SGLT1 on transcriptional and post-transcriptional levels, the nucleocytoplasmic shuttling of RS1 might ensure a rapid switch between short- and long-term regulations of SGLT1 by RS1. Whether confluence dependent nuclear location is required for the "cytoplasmic" function of RS1 is unclear and calls for further investigation.

Notably, RS1-NS overlaps with a recently identified transcription modulatory domain of RS1 located between amino acids 328-529 of hRS1 (C. Chintalapati, R. Poppe, V. Gorboulev and H. Koepsell, unpublished data). Thus, amino acids 328-529 represent a domain which consists of two overlapping functional subunits: a nuclear transport subunit and a transcription modulatory subunit. It is therefore possible that proteolytic degradation of RS1 leads not only to a relieve of TGN regulatory domain of RS1 (Korn *et al.*, 2001;Veyhl *et al.*, 2006;Kroiss *et al.*, 2006) but also to a release of a transcription regulating domain. Since RS1 was suggested to downregulate SGLT1 on the transcriptional level, nuclear location of RS1 might be important for the regulation of SGLT1. The nuclear location of RS1 suggests that RS1 can directly interact with the transcriptional complex of SGLT1. However, RS1 might also target an unknown factor, e.g., a transcription factor, which, in turn, regulates SGLT1.

In an attempt to characterize the targets of RS1 in the nucleus, we performed the gene expression profiling of mouse embryonic fibroblasts with RS1$^{-/-}$ genotype in comparison to wild-type fibroblasts (see Appendix I). Interestingly, our results suggest that transcriptional regulation by RS1 might be important for the cell cycle and cell division. Since RS1 localization depends on the cell cycle, it is tempting to speculate that RS1 might be important for the cell cycle progression or regulation during specific phases of the cell cycle.

Appendix I. Gene expression profiling in RS1 deficient mouse embryonic fibroblasts.

In an attempt to characterize the role of RS1 in the nucleus, we compared gene expression profiles of wild-type mouse embryonic fibroblasts (MEFs) and MEFs in which the gene encoding RS1 was disrupted (Osswald et al., 2005). The RS1$^{-/-}$ and wild-type embryos were derived from the same heterozygous RS1$^{+/-}$ animal, and three wild-type – knock-out pairs from three different mice were used. The obtained fibroblasts were cultivated until the passage number 3-4 and synchronized to S phase of the cell cycle. Thereafter, the cells were lysed, and the corresponding total RNAs were purified. Subsequently, the analysis of the quality of RNA, microarray analysis as well as normalization and evaluation of the raw data were performed in collaboration with Dr. Susanne Kneitz (Interdisciplinary Centre for Clinical Research, Institute of Immune Biology, University of Würzburg).

A statistical analysis of the 29012 genes detected 438 reproducibly differentially expressed genes between RS1$^{-/-}$ knock-out and wild-type fibroblasts (adjusted p-value < 0.01). We could not see any general change of the transcript abundance of glucose transporters of the GLUT and SGLT families (Table 7). These findings are in agreement with the previous findings which showed that mRNA expression levels of two major glucose transporters in the small intestine, SGLT1 and GLUT2, do not differ between the intestines of the wild-type and RS1$^{-/-}$ mice (Osswald et al., 2005). We also analysed expression of the genes encoding factors that have been shown to regulate SGLT1 on the transcriptional level. The mRNA levels of these genes was not changed. They include HNF1 homeobox B *Hnf1b* (Rhoads et al., 1998;Wood et al., 1999;Martin et al., 2000;Vayro et al., 2001;Kekuda et al., 2008;Balakrishnan et al., 2008), trans-acting transcription factor 1 *Sp1* (Martin et al., 2000;Tabatabai et al., 2005;Kekuda et al., 2008), forkhead box L1 *Foxl1* (Katz et al., 2004), CCAAT/enhancer binding protein alpha *Cebpa* (Oesterreicher et al., 1998), interleukin 6 *Il6* (Lee et al., 2007), GATA binding protein 5 *Gata5* (Balakrishnan et al., 2008), and caudal type homeobox 2 *Cdx2* (Balakrishnan et al., 2008) (Table 7). Although the expression level of SGLT1 and of the factors that have been shown to regulate SGLT1 mRNA expression was not changed in RS1$^{-/-}$ MEFs in comparison with wild-type MEFs, it is possible that the regulation of SGLT1 by RS1 is cell subtype-, tissue- or organ-specific.

Since RS1 is supposed to act as a transcription factor or as a regulator of transcription factors, we analysed the mRNA expression levels of the transcription factors. Expression of most of them was not changed; however, a slight but significant difference in RS1$^{-/-}$ MEFs in comparison with wild type MEFs was observed for the activating transcription factor 5 *Atf5* (ratio between the wild-type and knock-out 0.8; P value 0.02). Whether the regulation of *Atf5* by RS1 has physiological significance has to be elucidated.

Table 7. Analysis of the mRNA levels of genes encoding glucose transporters and genes encoding factors which regulate SGLT1 expression in RS1-/- knock-out fibroblasts in comparison with the wild-type fibroblasts. The ratios between the knock-out and wild-type (ratio ko/wt) and the corresponding P values are indicated.

Gene symbol	Gene name	Accession No.	Ratio ko/wt	P value
Genes encoding glucose transporters				
Slc5a1	Sglt1	NM_019810	1	1
Slc5a2	Sglt2	NM_133254	0.94	1
Slc5a9	Sglt4	NM_145551	1	0.5
Slc5a10	Sglt5	NM_001033227	1	0.98
Slc2a1	Glut1	NM_011400	0.93	0.24
Slc2a2	Glut2	NM_031197	1	0.4
Slc2a3	Glut3	NM_011401	1	0.46
Slc2a4	Glut4	NM_009204	1	0.55
Slc2a5	Glut5	NM_019741	1	0.61
Slc2a6	Glut6	NM_172659	1	0.47
Slc2a8	Glut8	NM_019488	1	0.94
Slc2a9	Glut9	NM_001102414	1	0.92
Slc2a10	Glut10	NM_130451	0.94	0.25
Slc2a12	Glut12	NM_178934	1	0.49
Genes encoding factors which regulate SGLT1 expression				
Hnf1b	HNF1 homeobox B	NM_009330	0.98	0.57
Sp1	trans-acting transcription factor 1	NM_013672	1.02	0.68
Foxl1	forkhead box L1	NM_008024	1.02	0.68
Cebpa	CCAAT/enhancer binding protein alpha	NM_007678	1.07	0.19
Il6	interleukin 6	NM_031168	0.87	0.11
Gata5	GATA binding protein 5	NM_008093	1.03	0.57
Cdx2	caudal type homeobox 2	NM_007673	1.03	0.71

The most differentially expressed genes were the down-regulated genes calponin *Cnn1*, thymic stromal lymphopoietin *Tslp*, gap junction protein β2 *Gjb2*, smooth muscle actin γ2 *Actg2*, eukaryotic translation initiation factor 2, subunit 3, structural gene Y-linked *Eif2s3y*, protein tyrosine phosphatase receptor type Z *Ptprz1*, and the up-regulated genes dipeptidylpeptidase 7 *Dpp7*, metallothionein-1 *Mt1*, mitochondrial galactosidase alpha *Gla*, UDP-N-acetylglucosamine pyrophosphorylase1-like1 *Uap1l1* (Table 8).

Gene Ontology project (Ashburner *et al.*, 2000) was used to describe gene products in terms of their associated biological processes, cellular components and molecular functions and allowed us to extract intrinsic functional information from hundreds of significantly differentially expressed genes. The genes that displayed significantly different transcript abundance in RS1 -/- fibroblasts compared to wild-type fibroblasts (p-value < 0.01) were classified in the unsupervised clustering into 685 biological processes, cellular components, and molecular functions. Functional group enrichment analysis was used to identify categories with a significant enrichment in the number of genes differentially expressed in RS1-/- knock-out and wild-type fibroblasts. Down-regulation of RS1 especially affected the following biological processes: cell organization and biogenesis, cell division, cell cycle, nitrogen compound metabolism, muscle contraction, cell growth, cell-substrate adhesion, and chromosome segregation. The cellular components involved were the extracellular matrix, the intracellular non-membrane-bound organelle, and the chromosomal part. The molecular functions included cytoskeletal protein binding, growth factor binding, protein complex binding, intramolecular oxidoreductase activity.

Table 8. Most down-regulated and up-regulated genes in RS1-/- knock-out fibroblasts in comparison with the wild-type fibroblasts. The ratios between the knock-out and wild-type (ratio ko/wt) and the corresponding P values are indicated.

Gene symbol	Gene name	Accession No.	Ratio ko/wt	P value
Down-regulated genes				
Cnn1	calponin	NM_009922	0.24	0.0004
Tslp	thymic stromal lymphopoietin	NM_021367	0.30	0.0038
Gjb2	gap junction protein	NM_008125	0.32	8.70E-05
Actg2	smooth muscle actin gamma 2	NM_009610	0.34	0.0014
Eif2s3y	eukaryotic translation initiation factor 2, subunit 3, structural gene Y-linked	NM_012011	0.34	0.0067
Ptprz1	protein tyrosine phosphatase receptor type	NM_001081306	0.35	0.0011
Up-regulated genes				
Uap1l1	UDP-N-acteylglucosamine pyrophosphorylase1-like 1	NM_001033293	2.18	0.0004
Gla	mitochondrial galactosidase alpha	NM_013463	2.09	0.0034
Mt1	metallothionein-1	NM_013602	2.04	0.0124
Dpp7	dipeptidylpeptidase 7	NM_031843	1.99	0.0055

Obviously, not all of these genes have to be direct targets of RS1, and the observed regulation of some genes can be a secondary effect. In order to detect the more subtle effects of RS1 deletion on gene expression patterns, further investigations are absolutely required. Nevertheless, our results clearly demonstrate that the loss of RS1 affects gene expression pattern of embryonic fibroblasts. In future experiments, the microarray results have to be validated with RT-PCR. Moreover, a proteomics-based approach will be employed to reveal whether the observed difference in gene expression is also valid on the protein expression level. Furthermore, the physiological relevance of the obtained RS1-mediated effects has to be elucidated.

Appendix II. Studies on the ubiquitination of pRS1.

Ubiquitination of proteins plays an important role in a wide range of cellular processes including cell cycle control and progression, signal transduction, chromosome structure maintenance, transcriptional regulation, endocytosis, organelle biogenesis, viral pathogenesis, and the stress response (Hershko and Ciechanover, 1998;Hershko, 2005). In most cases polyubiquitination of a substrate leads to a fast degradation of proteins by the 26S proteasome (Ciechanover *et al.*, 1984) whereas monoubiquitination or oligoubiquitination is responsible for the non-degradative pathways regulating protein stability, function, and intracellular localization and functions as a signaling device to establish protein-protein interactions with intracellular proteins. Some data indicate that RS1 can be regulated by ubiquitination. RS1 expression was dependent on the proteasome inhibitor MG-132 (Leyerer, 2007). It suggests that RS1 should be polyubiquitinated since most of the proteasome substrates are recognized via polyubiquitinated chains (Hershko *et al.*, 2000).

Studying ubiquitination of RS1 *in vivo*, we were aware of two major factors which might complicate the analysis of ubiquitination: the low steady-state levels of the ubiquitinated forms caused by degradation by the 26S proteasome and/or highly active deubiquitinating enzymes (Dubs) that remove ubiquitin units (Pickart and Cohen, 2004;Bloom and Pagano, 2005). Therefore, we tried to design the experiments in such a fashion that in spite of these factors ubiquitination was preserved. We performed immunoprecipitation of YFP-hRS1 with anti-GFP antibodies (which also recognize YFP) and analyzed the precipitates by immunoblotting with antibodies recognizing ubiquitin. HEK 293 cells were cotransfected with YFP-hRS1 and FLAG-tagged ubiquitin (FLAG-Ub). Overexpression of ubiquitin increases protein ubiquitination whereas FLAG tag facilitates detection of ubiquitination. HEK 293 cells transfected with FLAG-Ub alone served as control. Since many polyubiquitinated proteins are rapidly degraded by the proteasome, in the second set of experiments, HEK 293 cells were cotransfected with YFP-hRS1 and FLAG-tagged ubiquitin mutant in which lysine 48 was mutated to arginine (FLAG-Ub(Lys48Arg)). The mutated ubiquitin cannot form polyubiquitin chains via lysine 48 that prevents recognition of the ubiquitinated proteins by the proteasome and thereby leads to accumulation of the ubiquitinated protein in the cell simplifying ubiquitination detection (Willems *et al.*, 1996). Another factor which might interfere with detection of protein ubiquitination is the presence of Dubs which can remove the ubiquitin molecules from the protein of interest. This problem was avoided by the addition of the N-ethylmaleimide (NEM) which blocks the critical cysteine residue in the active site of Dubs (Hjerpe and Rodriguez, 2008). Since NEM does not specifically inactivate Dubs and might also affect 26S proteasome and recognition sites of ubiquitination on RS1, in some experiments the ubiquitin aldehyde was used to inhibit deubiquitination. HEK cells expressing the corresponding proteins were lysed, and the lysates were used for immunoprecipitation. Lysates and elution fractions were analyzed by immunoblotting with antibodies recognizing GFP to control purification efficiency or FLAG-tag to trace ubiquitination. Whereas purification was successful as confirmed by immunoblotting with anti-GFP antibodies (Figure 11), ubiquitinated proteins could not be detected in the elution fractions (Figure 11). However, the ubiquitinated proteins were obtained in the lysate and the flow-through that indicates that ubiquitination could be preserved during the lysis and purification. It suggests that RS1 is neither monoubiquitinated nor polyubiquitinated in HEK cells. The data obtained with immunoblotting were confirmed by mass spectrometry analysis. By this analysis, ubiquitination could not be detected in the obtained elution fractions (Koepsell H., Filatova A., Reinders Y., unpublished data). Importantly, our data do not exclude the possibility of RS1 ubiquitination. RS1 might be ubiquitinated under specific conditions which were not covered by our experiment. For example, RS1 ubiquitination might be cell type-specific. Alternatively, it is also possible that in spite of all precautions aimed to prevent the deubiquitination of RS1 we could not preserve the ubiquitination. In this case, an alternative approach can be applied to study RS1 ubiquitination. For example, purification of ubiquitinated proteins can be performed with Ni^{2+}-chelate affinity chromatography if ubiquitin or RS1 is fused to hexahistidine tag. The main advantage of this procedure over the others is that the lysate preparation and purification are performed under highly denaturing conditions (8M urea or 6M guanidinium) which limit Dubs activity and largely preserve the ubiquitination status during the entire procedure (Kaiser and Tagwerker, 2005;Hjerpe and Rodriguez, 2008).

The finding that expression of pRS1 (Leyerer, 2007) and hRS1 (data not shown) is dependent on the proteasome inhibitor MG-132 seems to contradict the absence of polyubiquitinated RS1. However, the proteasome can mediate degradation of few proteins that do not undergo ubiquitination, for example, the c-Jun protein (Jariel-Encontre et al., 1995), the cycline dependent kinase inhibitor $p21^{waf1/cip1}$ (Jin et al., 2003), the Rb tumor suppressor (Sdek et al., 2005), and the steroid receptor coactivator-3 (SRC3/AIB1) (Li et al., 2006). Accordingly, Kruppel-like zinc finger transcriptional factor KLF-5 can be degraded by the proteasome through the ubiquitin-dependent as well as ubiquitin-independent pathway (Chen et al., 2007). Ornithine decarboxylase (ODC) is degraded by the proteasome in ubiquitin independent manner; this process is mediated by antizyme AZ1 (Murakami et al., 1992). Recently, ODC and KLF-5 were identified as interacting partners of RS1 in yeast two-hybrid screen (Chintalapati C., Koepsell H., unpublished data). It raises the tempting assumption that the degradation of RS1 might be regulated by the same factors as the degradation of ODC and/or KLF-5. On the other hand, since MG-132 also inhibits calpain, RS1 degradation might depend on calpain and be independent of the proteasome (see Appendix III).

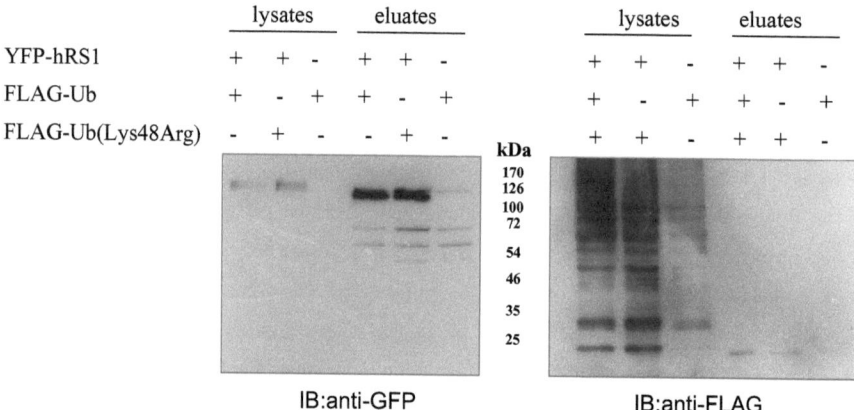

Figure 11. Studies on RS1 ubiquitination *in vivo*. HEK 293 cells were cotransfected with YFP-hRS1 and FLAG-Ub or FLAG-Ub(Lys48Arg). The HEK 293 cells transfected with FLAG-Ub served as control. Ubiquitination was assayed by immunoprecipitation of YFP-hRS1 with anti-GFP antibodies followed by immunoblotting. The purification efficiency was verified using anti-GFP antibodies, and the ubiquitinated proteins were detected with anti-FLAG antibodies.

Appendix III. Studies on the degradation of pRS1 in subconfluent and confluent LLC-PK1 cells. Investigation of the roles of the proteasome and calpain.

The activity of a protein depends on its concentration. The regulation of concentration of a protein occurs at many levels, including control of protein synthesis, especially at the level of transcription, and posttranslational modification, such as phosphorylation, which alters the concentration of active forms of the protein. Another widespread and effective form of posttranslational modification is proteolytic degradation, which can rapidly and irreversibly inactivate a protein by destroying it. Concentration of the pRS1 protein at the plasma membrane decreases dramatically when LLC-PK$_1$ cells reach confluence, and this effect is independent of transcription (Korn *et al.*, 2001). Interestingly, the decrease of pRS1 protein concentration in confluent cells is accompanied by a strong increase in SGLT1 protein expression (Korn *et al.*, 2001). Thus, the regulation of RS1 protein concentration might represent one of the mechanisms of regulation of SGLT1 expression. It has been observed previously that concentration of RS1 protein in LLC-PK$_1$ cells can be increased by the addition of the proteasome inhibitor MG-132 (Leyerer, 2007). Moreover, in confluent LLC-PK$_1$ cells, addition of MG-132 led to the increase of RS1 protein expression and redistribution of RS1 in the nucleus (Kroiss *et al.*, 2006). MG-132 is generally accepted as a highly potent inhibitor of the proteasome; however, this compound is not specific for the proteasome and also potently inhibits various cysteine proteases and calpains (Lee and Goldberg, 1998). In this study, we tried to investigate the effect of MG-132 on RS1 protein and to distinguish between the roles of the proteasome and calpain in degradation of RS1.

We analyzed the dependence of hRS1 protein concentration on the inhibitor MG-132 in HEK 293 cells (Figure 12). 40-50% confluent HEK 293 cells were transiently transfected with YFP-hRS1, hRS1-YFP, YFP-hRS1-FLAG-His$_8$, or FLAG-His$_8$-YFP-hRS1 and one day after transfection incubated with 10 μM MG-132 for 20 hours. The cells were lysed, and the lysates were analyzed by immunoblotting with anti-GFP antibodies. We observed that addition of a C-terminal tag (YFP or FLAG-His$_8$) to hRS1 leads to the stabilization of the protein in HEK 293 cells and abolishes the dependence of the protein concentration on MG-132 (Figure 12). How a C-terminal tag affects dependence of RS1 protein stability on MG-132 is not clear. One of the reasons might be the requirement of a free C-terminus that might be involved in interaction with proteins responsible for the stabilization/degradation of RS1. Alternatively, addition of the C-terminal tag could induce an alteration of the protein secondary structure which is essential for binding of the degrading enzyme(s) and/or regulatory proteins to RS1.

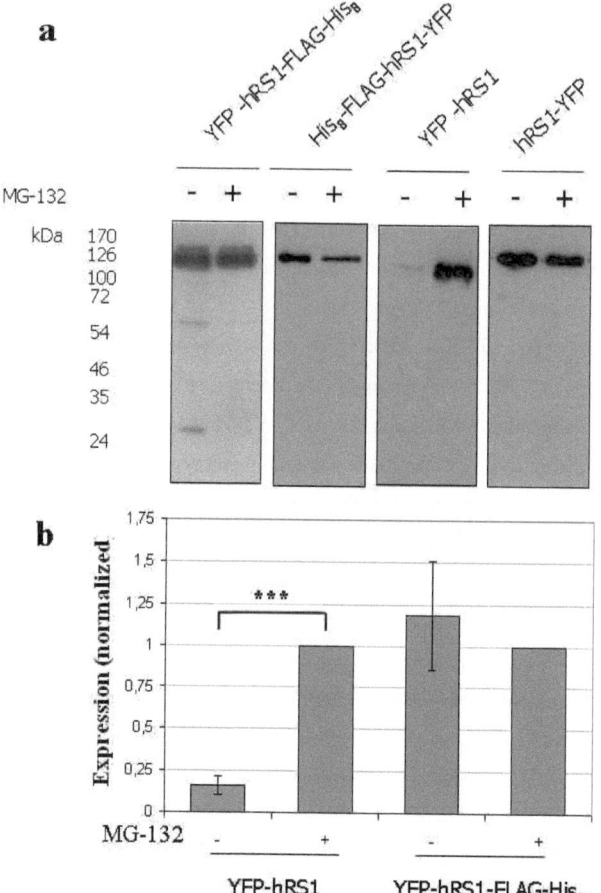

Figure 12. Analysis of the dependence of the protein stability on the proteasome inhibitor MG-132 for the proteins YFP-hRS1-FLAG-His$_8$, FLAG-His$_8$-YFP-hRS1, YFP-hRS1, and hRS1-YFP. *a.* HEK 293 cells were transiently transfected with an indicated construct, and one day after transfection the cells were incubated with a proteasome inhibitor 10 µM MG-132 for 20 hours. Thereafter, the cells were lysed and the lysates were analyzed by immunoblotting with anti-GFP antibodies. Per lane, 25 µg of total protein extract were applied. *b.* Relative expression level of YFP-hRS1 and YFP-hRS1-FLAG-His$_8$ in HEK 293 cells treated with MG-132 in comparison with non-treated cells. Densitometric quantification was perfrmed using programm Image J (see Methods). The data indicate that a C-terminal tag stabilizes hRS1 in HEK 293 cells and prevents its degradation by the proteasome.

Since addition of a C-terminal tag abolished dependence of hRS1 concentration on MG-132, we generated stable LLC-PK$_1$ cell lines which express GFP-pRS1 or GFP-pRS1-FLAG-His$_8$. By this mean we hoped to investigate the role of the degradation of RS1 in the regulation of SGLT1 on mRNA, protein and activity levels. Analysis of the protein expression in subconfluent and confluent LLC-PK$_1$ cells did not reveal any difference between the two proteins, and the protein expression of both proteins was significantly higher in subconfluent cells compared to confluent cells (Figure 13). To investigate the effect of MG-132, the cells were grown until 40-50% confluence or three days after confluence and incubated with or without 10 µM MG-132 for 20 hours. Thereafter, the cells were lysed, and the lysates were analyzed by immunoblotting with anti-GFP antibodies. Surprisingly, GFP-pRS1 and the C-terminally tagged GFP-pRS1-FLAG-His$_8$ did not differ in the response to MG-132. In confluent cells the concentration of both proteins was strongly dependent on MG-132 (Figure 14). The different effects of MG-132 on C-terminally tagged hRS1 and pRS1 in HEK 293 and confluent LLC-PK$_1$ cells, respectively, may be due to the RS1 ortholog- or cell line-specificity. This question was not investigated further. With anti-GFP antibodies, only the full-length pRS1 was detected. It indicates that the degradation of RS1 occurs without formation of relatively stable N-terminal fragments. The presence of C-terminal fragments of pRS1 was assessed using immunoblotting with anti-pRS1 antibodies (Figure 14). These antibodies were raised against the full-length pRS1, and the exact epitopes are not known (Valentin et al., 2000). Similarly to anti-GFP antibodies, with anti-pRS1 antibodies only the full length pRS1 was detected. It suggests that during pRS1 degradation C-terminal fragments are not formed; however, since recognition epitopes of the anti-pRS1 antibodies are not known, we cannot exclude the formation of small C-terminal fragments. In subconfluent LLC-PK$_1$ cells, the concentration of both GFP-pRS1 and GFP-pRS1-FLAG-His$_8$ was high and was not changed upon application of MG-132 (Figure 14).

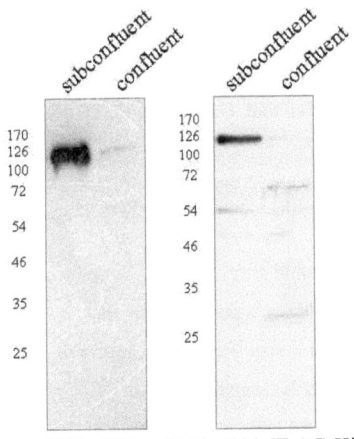

GFP-pRS1   GFP-pRS1-FLAG-His$_8$

Figure 13. Expression of GFP-pRS1 and GFP-pRS1-FLAG-His$_8$ in subconfluent and confluent LLC-PK$_1$ cells. *a*. LLC-PK$_1$ cells stably transfected with GFP-pRS1 or GFP-pRS1-FLAG-His$_8$ were grown until 70% confluence or four days after confluence and lysed. The lysates were analyzed by immunoblotting with anti-GFP antibodies. Per lane, 25 µg of total protein extract were applied.

The increase of the protein amount in confluent cells upon application of MG-132 suggests that the posttranscriptional downregulation of RS1 in confluent cells (Korn et al., 2001) may be due to the proteasome- or calpain-dependent degradation.

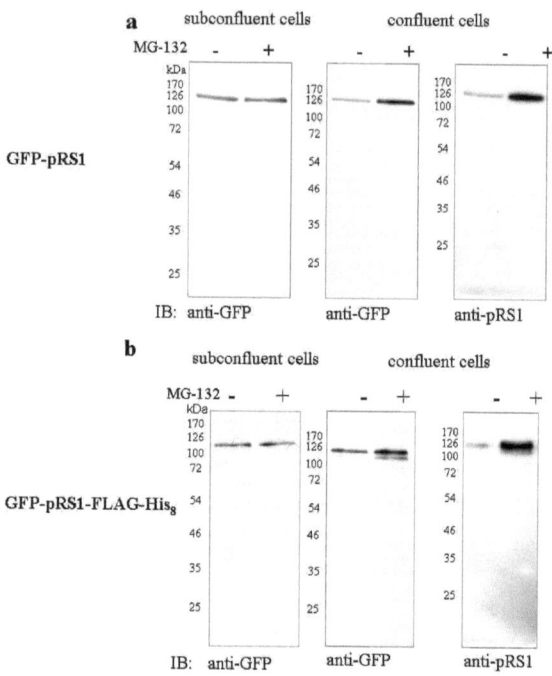

Figure 14. Dependence of GFP-pRS1 and GFP-pRS1-FLAG-His$_8$ expression in subconfluent and confluent LLC-PK$_1$ cells on MG-132. LLC-PK$_1$ cells stably transfected with GFP-pRS1 (*a*) or GFP-pRS1-FLAG-His$_8$ (*b*) were grown until 40-50% confluence or two days after confluence and incubated with or without MG-132 for 20 hours. Thereafter, the cells were lysed, and the lysates were analyzed by immunoblotting with anti-GFP or anti-pRS1 antibodies. Per lane, 25 µg of total protein extract were applied.

As MG-132 inhibits proteasome and calpain, we tried to distinguish between the effects of the proteasome and calpain on the RS1 expression. For that, a specific proteasome inhibitor MG 262 (Lee and Goldberg, 1998) and a specific calpain inhibitor calpeptin (Tsujinaka et al., 1988) were employed. LLC-PK$_1$ cells stably transfected with GFP-pRS1 were grown three days after confluence and incubated with 10 µM MG-132, 0.2 nM MG-262 or 50 µM calpeptin for 20 hours. Non-treated cells served as control. The cells were lysed, and the lysates were analyzed by immunoblotting with anti-GFP or anti-pRS1 antibodies. The amount of RS1 protein was increased in the cells that were incubated with MG-132 or the calpain inhibitor calpeptin. A slight increase was also observed for the cells incubated with the proteasome inhibitor MG-262 (Figure 15*a,b*).

The data indicate that RS1 concentration depends on calpain whereas the proteasome plays only a minor role. Notably, similarly to other calpain substrates, RS1 degradation products could not be identified (see Figures 12-14, 15a,b). These fragments are probably cleared rapidly in cells by endopeptidases or the proteasome thereby evading detection (Botbol and Scornik, 1983;Han *et al.*, 1999;Chen *et al.*, 2007).

In paired t-test, the differences between the amounts of GFP-pRS1 in calpeptin- or MG-262-treated cells in comparison with non-treated cells were significant (0.035 and 0.042, respectively). However, large variations between the single experiments were observed (Figure 15c,d). It might have different reasons. First, the expression of the proteins varied between the single experiments. In cells expressing higher amounts of the protein, the protein concentration might have been underestimated due to saturation of ECL reaction. Second, calpain and proteasome might be regulated by the factors which were not controlled in the present set of experiments. Further investigation is required to reveal the determinants of calpain- and proteasome-mediated cleavage. Notably, the effect of MG-132 correlated with the effects of calpeptin and MG-262, and in a single experiment either a strong or a slight increase upon application of the three inhibitors was observed. Calpains are cytoplasmic Ca^{2+}-dependent cysteine proteases which are localised at the plasma membrane, Golgi network and in the nucleus (Bevers and Neumar, 2008). Fifteen isoforms of calpain have been described; the majority of them have been identified only as mRNA, and several are thought to be tissue specific (Goll *et al.*, 2003). The ubiquitously expressed micromolar and millimolar Ca^{2+}-requiring neutral proteases (μ-calpain and m-calpain) are the best studied members of the family (Suzuki *et al.*, 2004;Bevers and Neumar, 2008). *In vitro*, m-calpain binds Ca^{2+} with relatively low affinity (millimolar), and μ-calpain binds with higher affinity (micromolar); however, their Ca^{2+} requirements in cells and tissues are influenced by several factors that may lower these requirements. Moreover, it was observed that calpain activation can occur without changes in intracellular calcium (Goll *et al.*, 2003). Both isoforms are thought to have indistinguishable substrate affinities (Bevers and Neumar, 2008). Calpains have diverse functions catalyzing the proteolysis of proteins involved in cytoskeletal remodelling, cell cycle regulation, signal transduction, cell differentiation, apoptosis and necrosis, embryonic development, and vesicular trafficking (Zatz and Starling, 2005;Bevers and Neumar, 2008). Interestingly, calpains can shuttle between the nucleus and the cytoplasm and have distinct functions in proliferating and differentiated cells (Tremper-Wells and Vallano, 2005). Moreover, calpains are supposed to play a role in intestinal differentiation (Ibrahim *et al.*, 1994;Potter *et al.*, 2003). It reinforces the assumption that the degradation of RS1 might be important for the regulation of SGLT1 during confluence of LLC-PK_1 cells.

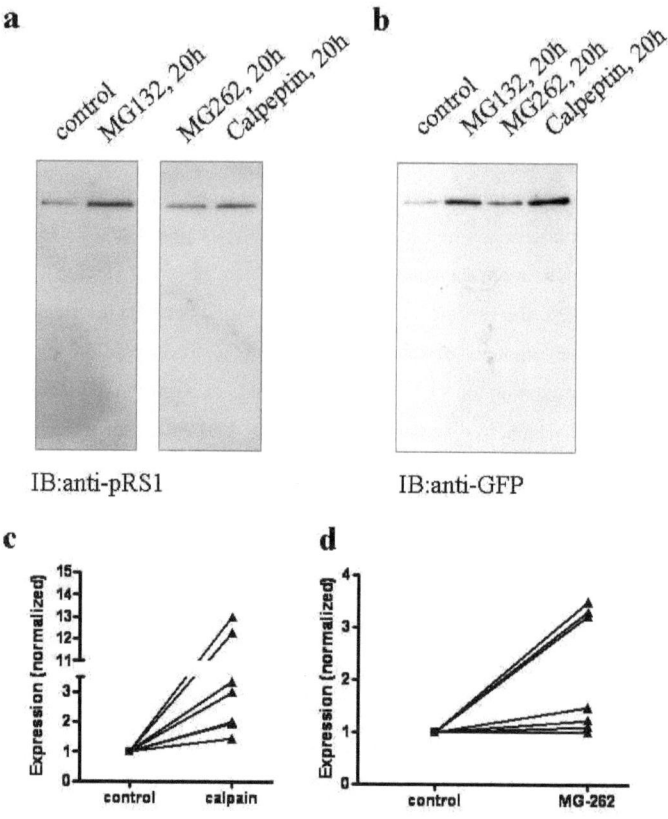

Figure 15. Effects of MG-132, MG-262 and calpeptin on the concentration of GFP-pRS1 protein in subconfluent LLC-PK$_1$ cells. LLC-PK$_1$ cells stably transfected with GFP-pRS1 were grown three days after confluence and incubated with 10 µM MG-132, 0.2 nM MG-262, or 50 µM calpeptin for 20 hours. Non-treated cells served as control. Thereafter, the cells were lysed, and the lysates were analyzed by Western blot with anti-pRS1 (*a*) or anti-GFP (*b*) antibodies. Per lane, 25 µg of total protein extract were applied. (*c,d*) Normalized expression of GFP-pRS1 in calpain- (*c*) or MG-262- (*d*) treated cells in comparison with non-treated cells.

Some properties of a protease that performs cleavage of RS1 have been identified in experimental work with RS1. First, during purification, the RS1 protein was stabilised by the addition of EDTA indicating that the protease performing cleavage of RS1 is Ca^{2+} or Mg^{2+} dependent (Thorsten Keller, personal communications). Second, the concentration of pRS1 protein in LLC-PK$_1$ cells did not depend on lysosome (Leyerer, 2007); therefore, the protease should be cytoplasmic. Moreover, according to RS1 localization, the protease should be localized in the nucleus or at the *trans*-Golgi network. Third, since the confluence dependent downregulation of RS1 was observed in porcine kidney LLC-PK$_1$ cells, the protease should be expressed in the kidney. Strikingly, calpain fits all these criteria.

We tried to identify putative calpain cleavage sites and recognition motifs. Targets of calpains do not possess common recognition motif(s) (Friedrich and Bozoky, 2005). Despite multiple attempts of substrate sequence analysis (Tompa *et al.*, 2004;Cuerrier *et al.*, 2005), no definitive method exists to predict whether a given compound is a calpain substrate, or, if it is a substrate, to identify the cleavage site. Using naturally occurring and artificial peptides as substrates for calpain, Sasaki et al. (1984) formulated a general preference rule for calpain proteolysis at the cleavage site: a Lys, Tyr, Arg or Met residue in the P1 position preceded by a hydrophobic amino acid residue (Leu or Val) in the P2 position would favour cleavage at the carboxyl side of the residue in the P1 position:

$$\downarrow$$

P2(L/V)-P1(K/Y/R/M)-

However, this pattern does not explain all the cases. Some proteins containing these motifs are not substrates for calpain and *visa versa* some proteins that do not possess these motifs are degraded by calpain (Wang *et al.*, 1989). Nevertheless, we analyzed RS1 sequences from human, pig, rabbit and mouse for the presence of the putative calpain cleavage sites. The analysis revealed the presence of multiple potential calpain cleavage sites according to the preference rule (Figure 16, 17b).

It is supposed that prior to cleavage calpain attaches to specific recognition sequences on a substrate. The recognition sequences are suggested to localize adjacent to the actual cleavage site and to be represented by a higher order structure(s) in substrate (Rechsteiner and Rogers, 1996). Two major calpain recognition motifs are known: Proline-Glutamate-Serine-Threonine (PEST) motifs (Rechsteiner and Rogers, 1996) and/or calmodulin binding sites (Wang *et al.*, 1989).

PEST motifs are the sequences enriched in proline, glutamate, serine and threonine flanked by positive amino acids. They are present in about 95% of rapidly turning over proteins in eukaryotes and have been shown to trigger rapid intracellular degradation (Rechsteiner and Rogers, 1996). PEST motifs have been shown to mediate calpain-dependent (Wang *et al.*, 1989;Rechsteiner and Rogers, 1996;Shumway *et al.*, 1999) as well as proteasome-dependent degradation (Roth *et al.*, 1998). It is proposed that PEST motifs increase the local calcium concentration and, in turn, activate calpain (Sandoval *et al.*, 2006). PEST-FIND program (Rogers *et al.*, 1986) was used to identify putative PEST motifs in RS1 orthologs. This program predicts putative PEST motifs with scores ranging from -45 to +50. Scores more than 5.0 are considered significant and indicate strong PEST signals whereas scores between 0 and 5.0 indicate weak PEST signals. One to four strong PEST signals were identified in each RS1 ortholog (Figure 16, 17a); however, they were not conserved between the ortholog sequences.

Next, search of the putative calmodulin binding sites in hRS1 and pRS1 was performed employing MiniMotif Miner (Balla *et al.*, 2006) and Calmodulin Target Database (Yap *et al.*, 2000) (http://calcium.uhnres.utoronto.ca/ctdb/ctdb/home.html). Surprisingly, the two programs predicted different potential calmodulin binding sites (Figure 16, 17b). The Calmodulin Target Database predicted the existence of an "unclassified" calmodulin binding site at the N-terminus of RS1 which is conserved between the human and porcine orthologs. MiniMotif Miner predicted seven and eight potential calmodulin binding sites in hRS1 and pRS1, respectively (Figure 16, 17b). Several of them were conserved between the orthologs whereas others were species-specific.

The identification of a minimal "degron" that mediates degradation of RS1 would be a challenging task for future investigations. It would be particularly interesting to assess the roles of the PEST sequences and of other potential calpain recognition motifs in the regulation of degradation and function of RS1.

```
pRS1    MSSLPTSDGFNHQAHPSGQRPEIGSPPSLAHSVSASVCPFKPSDPDSIEPKAVKAVKALK
hRS1    MSSLPTSDGFNHPARSSGQSPDVGNPMSLARSVSASVCPIKPSDSDRIEP---KAVKALK
rbRS1   MSSSPPLDGSDHPAHSSGQSPEAGNPTSLARSVSASVCPVKPDNPDSTEP---EAVTALE
mRS1    MSSLPTSDGFDHPAP-SGQSPEVGSPTSLARSVSASVCAIKPGDPNSIES---LAMEATK
        ***.*.**.:*.*   ***.*:.*.* ***:******..**.:.:  *.    *:.*.:

pRS1    ASAEFQITFERKEQLPLQDPSDCASSADNAPANQTPAIPLQNSLEEAIVADNLEKSAEGS
hRS1    ASAEFQLNSEKKEHLSLQDLSDHASSADHAPTDQSPAMPMQNSSEEITVAGNLEKSAERS
rbRS1   ASDGFQINSKQTDRLPLQGHSPCAAAAAPS-----SAMPLRHSSEAAGVADSLEASAERR
mRS1    ASAEFQTNSKKTDPPPLQVLPDLASSAEQS-----LAMPFHKSSKEAVVAGNLEKSVEKG
        **  **  .  ::.:  .**   . *::*  :     *:*::::*  :   **..**.*.*

pRS1    TQGLKSHLHTRQEASLSVTTTRMQEPQRLIGEKGWHPEYQDPSQVNGLQQHEEPRNEQHE
hRS1    TQGLKFHLHTRQEASLSVTSTRMHEPQMFLGEKDWHPENQNLSQVSDPQQHEEPGNEQYE
rbRS1   TQGLRFHLHTRQEVNLSITTTRMHEPQMFAGEEGWHPENQNPSQVNDLQQHQEPENARHE
mRS1    TQGLRVYLHTRQDASLTLTTTGMREPQIFAEEKSWHPENQTPSPVNGLQQHRETGSVQRE
        ****: :*****:..*:.*:*  *:***   *:.****  *  *..***.*..  .: *

pRS1    VVQQNAPHDPEHLCNTGDLELLGERQQNQPKSVGLETAVRGDRPQQDVDLPGTEKNILPY
hRS1    VAQQKASHDQEYLCNIGDLELPEERQQNQHKIVDLEATMKGNGLPQNVDPPSAKKSIPSS
rbRS1   AGPRDAPSD------TGDLELPGERQQ-KHEVADREATMRGGRLQQDAGLPDPGKGALPS
mRS1    AGQQSVPQDQGCLCDAEDLELHEEVVS--------LEALRKGELQRHAHLPSAEKGLPAS
        .  :...*      ****  .    :::  .   :...*..*. .

pRS1    GCPGCGSSETFMEIDTVEQSLVAVLNSAGGQNTSVRNISASDLTVDNPLMEVETLKCNPS
hRS1    ECSGCSNSETFMEIDTAQQSLVTLLNSTGRQNANVKNIGALDLTLDNPLMEVETSKCNPS
rbRS1   GHCGRPDSETLMEVDAAEQSLVAVLSSS------VGNGSASGLTLGNPLMEVELPTCSPS
mRS1    GLCSCPCSEALMEVDTAEQSLVAMCSSTGRQDAVIKSPSVAHLASDNPTMEVETLQSNPS
        .  .**::**:*:.:.:*****..:.*      :     . .:. .** ****  .:**

pRS1    SEFLSNPTSTQNLQLPESSVEMSGTNKEYGNHPSSLSLCGTCQPSVESAEESCSSITAAL
hRS1    SEILNDSISTQDLQPPETNVEIPGTNKEYG-HYSSPSLCGSCQPSVESAEESCPSITAAL
rbRS1   SEILNGSIPIQDLQPPEGSVEMPGTDRAYGGRASSSSVCGSSQPPAESAEESCSSITTAL
mRS1    CEPVEHSILTRELQLPEDNVDMSTMDNKD--DNSSSLLSGHGQPSVESAEEFCSSVTVAL
        .*.: .    ::**.**.*.:::. :.        **  :.*  **..***** *.:*.**
```

81

```
pRS1    KELHELLVISSKPALENTSEEVTCRSEIVTEGQTDVKDLSERWTQSEHLTAAQNEQCSQV
hRS1    KELHELLVVSSKPASENTSEEVICQSETIAEGQTSIKDLSERWTQNEHLTQN--EQCPQV
rbRS1   KELHELLVISSKPASEAAYEEVTCQSEGTAWGQTRVNPS-ERWTESERRTQDE----DRP
mRS1    KELHELLVISCKPASEESPEHVTCQSEIGAESQPSVSDLSGRRVQSVHLTPSD--QYSQG
        ********:*.*** * : *.* *:**   : .*. :.    * .:.: *          :

pRS1    SFYQATSVSVKTEELTDTSTDAGTEDVENITSSGPGDGLLVDKENVPRSRESVNESSLVT
hRS1    SFHQAISVSVETEKLTGTSSDTGREAVENVNFRSLGDGLSTDKEGVPKSRESINKNRSVT
rbRS1   QVSHAIPECVKTEKLTDASPDTRIEDGENATFQGPGGGLSTDHG-APRSRGSVHESRSVT
mRS1    SCHQATSESGKTEIVG-TAPCAAVEDEASTSFEGLGDGLSPDREDVRRSTESARKSCSVA
        . :*. . :** : ::. :  *  .. .*.**  *: . :* * .:. *:

pRS1    LDSAKTSNQPHCTLGVEISPGLLAGEEGALNQTSEQTESLSSSFILVKDLGQGTQNPVTN
hRS1    VTSAKTSNQLHCTLGVEISPKLLAGEEDALNQTSEQTKSLSSNFILVKDLGQGIQNSVTD
rbRS1   VTSAETSNQSHRTLGVEISPRLLTGEGDALSQTCEQTKSL-----LVKDLGQGTQNPAPD
mRS1    ITSAKLSEQLPCTSGVEIAPELAASEG----------------AHSQPSEHVHNPGPD
        : **: *:*  * ****:* * :.*                    .: .: :*. .:

pRS1    RPETRENVCPEAAGLRQEFEPPTSHPSSSPSFLAPLIFPAADIDRILRAGFTLQEALGAL
hRS1    RPETRENVCPDASRPLLEYEPPTSHPSSSPAILPPLIFPATDIDRILRAGFTLQEALGAL
rbRS1   RPATREDVCRDAARPSLEVEAPPSHSSG-PCILPPLGFPAADIDRILRAGFTLQEALGAL
mRS1    RPETSS-VCPGAGLPRSGLDQPPTQSLSTPSVLPPFIFPAADVDRILGAGFTLQEALGAL
        ** *  . ** *.        : *.::. . *..*.*: ***:*:**** ************

pRS1    HRVGGNADLALLVLLAKNIVVPT
hRS1    HRVGGNADLALLVLLAKNIVVPT
rbRS1   HRVGGNADLALLVLLAKNIVVPT
mRS1    HRVGGNADLALLVLLAKNIVVPT
        ***********************
```

Figure 16. Potential calpain recognition motifs and cleavage sites. Light grey shadings indicate potential calpain cleavage sites; PEST motifs are shown in bold face. Asterisks and colons indicate identical and similar amino acids, respectively. Alignment of sequences of RS1 orthologs from porcine RS1 (pRS1), human RS1 (hRS1), rabbit RS1 (rbRS1) and mouse RS1 (mRS1) was performed with Clustal W (Version 1.83) (see Methods).

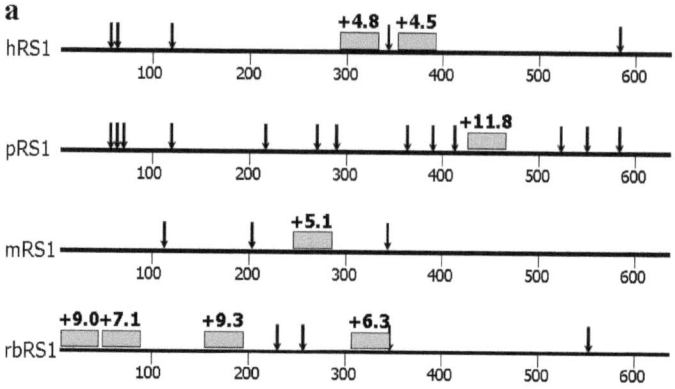
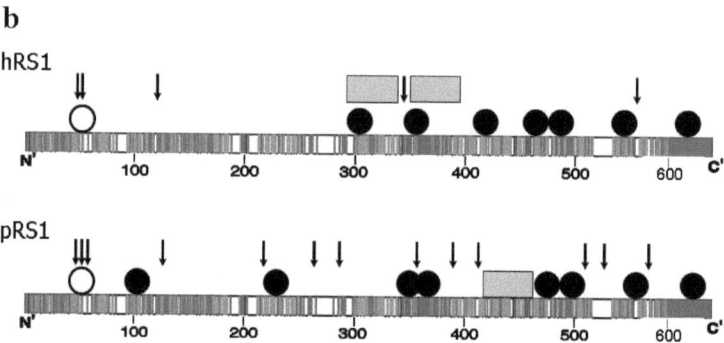

Figure 17. Schematic representation of PEST motifs in RS1 orthologs (*a*) and potential calpain cleavage sites and recognition motifs in hRS1 and pRS1 (*b*). The potential calpain recognition motifs include PEST motifs and calmodulin binding sites. The grey rectangles represent PEST motifs. Calmodulin binding sites predicted by calmodulin target database or MiniMotif Miner are indicated by white or black circles, respectively. The arrows indicate potential calpain cleavage sites. In *a*, the PEST scores are shown above the indicators of PEST motifs.

Appendix IV. Generation of antibodies which recognize phosphorylated serine 370.

In parallel with mass spectrometry analysis, immunodetection of the phosphorylated proteins was applied to investigate possible phosphorylation events of pRS1 *in vivo*. To this end, we aimed to generate antibodies which specifically recognize phosphorylated serine 370 (anti-P-Ser370). The potential specificity of such antibodies, generated against specific phosphorylated epitopes, makes site-specific recognition feasible, provided the relevant antibodies can be obtained.

For generation of anti-P-Ser370 antibodies, a peptide containing phosphorylated serine 370, ELHELLVIpSSKPALENTSC, was conjugated to the carrier protein ovalbumin and used for the immunization of two rabbits. Sera were collected from the animals every 3 weeks starting on day 35 after the initial immunization. For characterization of the obtained samples, enzyme linked immunosorbent assay (ELISA) tests were performed. The titers against phosphorylated peptide and non-phosphopeptide (cross-reaction) were determined (Table 9). The phosphorylated peptide was recognized by all sera better than the nonphosphorylated peptide (Table 9). The serum with the highest ratio between the titers against phosphorylated and nonphosphorylated peptides (full serum 149 I) was used for the antibody purification.

For purification of the antibodies, a two-step protocol was applied. First, the affinity purification with the nonphosphorylated peptide as an antigen was carried out to separate the antibodies recognizing the phosphorylated epitope from the non-phosphospecific antibodies. Subsequently, the flow-through was used for purification on phosphopeptide column to obtain exclusively phosphospecific antibodies separated from antibodies to other epitopes. The efficiency of the purification and the phosphospecificity of the antibodies were controlled by ELISA. As shown in the Table 9, the flow-through after the first purification step did not contain non-phosphospecific antibodies indicating that all antibodies which recognize non-phosphopeptide were efficiently removed. At the same time, the flow-through contained antibodies against the phosphopeptide. The titer in the flow-through was reduced in comparison with the full serum that can be explained by the non-specific binding of the antibodies to the non-phosphopeptide column or cross-reaction. After the second purification step, a high titer of antibodies recognizing the phosphopeptide (Table 9) indicated efficient purification of the antibodies. These antibodies virtually did not recognize non-phosphorylated peptide and, thus, represent phosphospecific antibodies.

Remarkably, one-step affinity purification on phosphopeptide column was not sufficient for purification of phosphospecific antibodies. The antibodies obtained according to this procedure recognized both non-phosphorylated and phosphopeptides (Table 9). The most probable reason is that the antibodies contained in the sera recognize not only the phosphoserine 370 but also other epitopes besides of the phosphorylation site.

Table 9. Titers and cross reaction of immunized rabbit sera and purified antibodies determined by ELISA. For full sera, number of immunized rabbit (according to the laboratory nomenclature) and number of bleed (I-IV) are indicated.

Serum/Antibody	Titer (anti-phosphopeptide)	Cross-reaction (anti-non-phosphorylated peptide)
Full sera		
Full serum 149 I	1 : 150 000	1 : 42 000
Full serum 149 II	1 : 205 000	1 : 80 000
Full serum 149 IV	1 : 150 000	1 : 60 000
Full serum 150 I	1 : 78 000	1 : 50 000
Full serum 150 IV	1 : 300 000	1 : 135 000
Two-step purification		
Flow-through after purification with non-phosphopeptide column	1 : 55 000	---
Purified Antibodies	1 : 50 000	1 : 700
One-step purification		
Purified antibodies	1 : 200 000	1 : 51 000

To assess the phosphorylation status of serine 370 in subconfluent and confluent cells, the affinity purification followed by immunoblotting with the generated phosphospecific antibodies was performed. LLC-PK$_1$ cells were transiently transfected with GFP-CK2-NS-PKC-PKC-β-Gal, and two days after transfection GFP fusion proteins were purified with anti-GFP antibodies from subconfluent and confluent LLC-PK$_1$ cells. To prevent dephosphorylation or phosphorylation of serine 370 during lysis and purification, all purification steps were performed on ice and all buffers were supplemented with phosphatase and kinase inhibitors. The efficiency of purification was controlled by the silver staining of polyacrylamide gels after electrophoresis and immunoblotting with anti-GFP antibody. The presence of the phosphorylated serine 370 was examined by immunoblotting with anti-phosphopeptide antibodies (Figure 18a). Whereas βGal-CK2-NS-PKC-PKC-GFP was detected in both lysate and eluates with anti GFP antibodies (Figure 18a), in a series of experiments we were unable to detect phosphorylation neither in subconfluent nor in confluent cells using the anti-phosphoserine 370 antibodies (Figure 18b). At the same time, the antibodies recognised various proteins and/or protein fragments in the lysate (Figure 18b). It might reflect either the recognition of the endogenous pRS1 and/or its fragments or binding of antibodies to proteins unrelated to pRS1. In the latter case, the antibodies reaction might represent either recognition of other phosphorylated proteins by the antibodies or non-specific binding of antibodies.

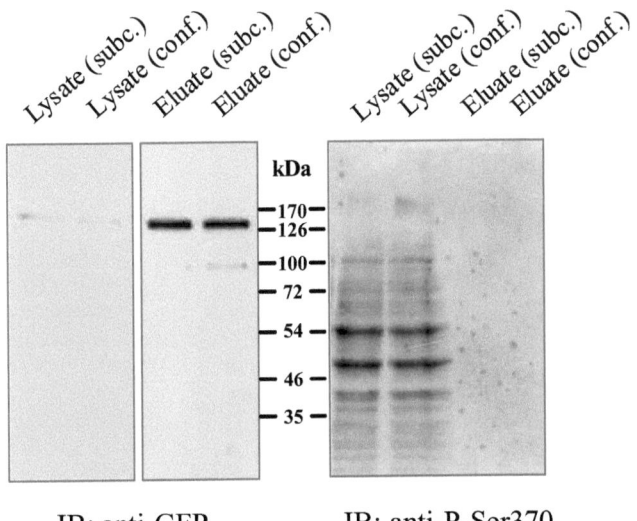

IB: anti-GFP   IB: anti-P-Ser370

Figure 18. Studies on the phosphorylation of RS1 *in vivo* using anti-P-Ser 370 antibodies. LLC-PK$_1$ cells were transiently transfected with GFP-CK2-NS-PKC-PKC-β-Gal. After immunoprecipitation with anti-GFP antibodies, the samples were divided into two parts for immunoblotting with anti-GFP antibodies or anti-phosphoserine 370 antibodies (anti-P-S370).

The absence of binding of the antibodies recognizing the phosphopeptide to βGal-CK2-NS-PKC-PKC-GFP might be due to different reasons. First, the purified βGal-CK2-NS-PKC-PKC-GFP might be not phosphorylated at serine 370. However, mass spectrometric analysis shows that the protein is phosphorylated. Second, the antibodies might be not phosphospecific against the serine 370 (see above). Third, the antibodies might not be suitable for the immunoblotting. For example, the recognition epitopes might be hidden due to refolding of the protein on the PVDF membrane (Birk and Koepsell, 1987). Although the generated antibodies could not be successfully used for the immunoblotting, the results of ELISA measurements suggest that the obtained antisera contain a pool of antibodies recognizing phosphoserine 370. Therefore, it might be worth testing all sera with immunoblotting. There can be a serum which recognizes phosphorylated serine 370 but does not cross-react with other proteins. The sera with a ratio between the titers against phospho- and non-phosphorylated peptides lower than in the serum 149I, that was used for the antibodies purification, still can be more RS1-specific. The higher cross-reaction can be overcome by two-step affinity purification which efficiently separates the phosphospecific antibodies from the antibodies recognizing non-phosphorylated peptide.

6. Summary

The RS1 protein (gene *RSC1A1*) participates in regulation of Na^+-D-glucose cotransporter SGLT1 and some other solute carriers. In subconfluent LLC-PK_1 cells, RS1 inhibits release of SGLT1 from the *trans*-Golgi network and transcription of SGLT1. In subconfluent cells, RS1 is localized in the nucleus and the cytoplasm whereas confluent cells contain predominantly cytoplasmic RS1.

In the present study, the mechanism and regulation of confluence-dependent nuclear location of RS1 was investigated. Confluence dependent nuclear location of RS1 was shown to be regulated by the cell cycle. A nuclear shuttling signal (NS) in pRS1 was identified that ensures confluence-dependent distribution of pRS1 and comprises nuclear localization signal (NLS) and nuclear export signal (NES). The NLS and NES of RS1 mediate translocation into and out of the nucleus via importin ß1 and CRM1, respectively, and the nuclear/cytoplasmic distribution of the RS1 protein is determined by the nuclear export activity. The adjacent protein kinase C (PKC) phosphorylation site at serine 370 of pRS1 was shown to control nuclear localization driven by NS and is necessary for the differential localization of RS1 in quiescent versus proliferating cells. Basing on the data of site-directed mutagenesis, PKC activation experiments and mass spectrometry analysis of RS1 phosphorylation, the following model of the regulation of RS1 nuclear location in LLC-PK_1 cells was proposed. In subconfluent cells, RS1 is actively imported into the nucleus whereas nuclear export of RS1 is not active leading to accumulation of RS1 in the nucleus. After confluence, phosphorylation of serine 370 of pRS1 by PKC takes place leading to enhancement of RS1 nuclear export and predominantly cytoplasmic distribution of the protein in the confluent cells.

The confluence-dependent regulation of RS1 localization may control SGLT1 expression during regeneration of enterocytes in small intestine and during regeneration of renal tubular cells after hypoxemic stress. Moreover, the gene expression profiling of mouse embryonic fibroblasts with $RS1^{-/-}$ genotype suggests that transcriptional regulation by RS1 might be important for the cell cycle and cell division. Since RS1 localization depends on the cell cycle, RS1 might play a role in the regulation of the solute carriers during specific phases of the cell cycle.

7. Abbreviations

aa	amino acid
AMG	methyl-α-D-glucopyranoside
APS	ammonium persulfate
DMEM	Dulbecco's modified Eagle's medium
DMSO	dimethyl sulfoxide
dNTP	deoxynucleotide triphosphate
DRB	5,6-dichlorobenzimidazole riboside
ECL	enhanced chemiluminescence
FCS	fetal calf serum
GFP	green fluorescent protein
HEK	human embryonic kidney
HRP	horse radish peroxidase
IRIP	ischemia/reperfusion inducible protein
Kap	karyopherin
LMB	leptomycin B
MEF	mouse embryonic fibroblast
NE	nuclear envelope
NEM	N-ethylmaleimide
NES	nuclear export signal
NLS	nuclear localization signal
NPC	nuclear pore complex
NSS	nuclear shuttling signal
OAT	organic anion transporter
OCT	organic cation transporter
PBS	phosphate buffered saline
PCR	polymerase chain reaction
PEI	polyethylenimine
PMA	phorbol 12-myristate 13-acetate
pRS1-NES	nuclear export signal of pRS1, aa 360-368
pRS1-NRS	nuclear localization signal of pRS1, aa 349-369
pRS1-NSS	nuclear shuttling signal of pRS1, aa 349-369
PVDF	polyvinylidene difluoride
RT	room temperature
SGLT	Na+-d-glucose cotransporter
STEBP	sterol regulatory element binding protein
TEMED	N,N,N',N'-Tetramethylethylendiamine
TEV	tobacco etch virus
TGN	trans-Golgi netwotk
YFP	yellow fluorescent protein

8. References

Afroze,T. and Husain,M. (2001). Cell cycle dependent regulation of intracellular calcium concentration in vascular smooth muscle cells: a potential target for drug therapy. Curr. Drug Targets. Cardiovasc. Haematol. Disord. *1*, 23-40.

Amsler,K. (1994). Role of cell density/cell-cell contact, and growth state in expression of differentiated properties by the LLC-PK$_1$ cell. J. Cell Physiol. *159*, 331-339.

Amsler,K. and Cook,J.S. (1982). Development of Na+-dependent hexose transport in a cultured line of porcine kidney cells. Am. J. Physiol. *242*, C94-C101.

Amsler,K., Ghatani,S., and Hemmings,B.A. (1991). cAMP-dependent protein kinase regulates renal epithelial cell properties. Am. J Physiol *260*, C1290-C1299.

Areces,L.B., Matafora,V., and Bachi,A. (2004). Analysis of protein phosphorylation by mass spectrometry. Eur. J. Mass Spectrom. (Chichester, Eng) *10*, 383-392.

Arts,G.J., Fornerod,M., and Mattaj,I.W. (1998). Identification of a nuclear export receptor for tRNA. Curr. Biol. *8*, 305-314.

Ashburner,M. *et al.* (2000). Gene ontology: tool for the unification of biology. The Gene Ontology Consortium. Nat. Genet. *25*, 25-29.

Asscher,Y., Pleban,S., Ben-Shushan,M., Levin-Khalifa,M., Yao,Z., and Seger,R. (2001). Leptomycin B: An Important Tool for the Study of Nuclear Export. In: LifeScience, A technical application bulletin.

Bachmann,R.A., Kim,J.H., Wu,A.L., Park,I.H., and Chen,J. (2006). A nuclear transport signal in mammalian target of rapamycin is critical for its cytoplasmic signaling to S6 kinase 1. J. Biol. Chem. *281*, 7357-7363.

Balakrishnan,A., Stearns,A.T., Rhoads,D.B., Ashley,S.W., and Tavakkolizadeh,A. (2008). Defining the transcriptional regulation of the intestinal sodium-glucose cotransporter using RNA-interference mediated gene silencing. Surgery *144*, 168-173.

Balla,S. *et al.* (2006). Minimotif Miner: a tool for investigating protein function. Nat. Methods *3*, 175-177.

Barret,A.J., Rawlings,N.D., and Woessner,J.F. (2004). Trypsin. Handbook of Proteolytic Enzymes. Elsevier Academic Press: San Diego, CA, 1483-1488.

Beguin,P., Mahalakshmi,R.N., Nagashima,K., Cher,D.H., Kuwamura,N., Yamada,Y., Seino,Y., and Hunziker,W. (2005). Roles of 14-3-3 and calmodulin binding in subcellular localization and function of the small G-protein Rem2. Biochem. J. *390*, 67-75.

Bevers,M.B. and Neumar,R.W. (2008). Mechanistic role of calpains in postischemic neurodegeneration. J. Cereb. Blood Flow Metab *28*, 655-673.

Bier ME (2002). Analysis of Proteins by MS. In: *Modern Protein Chemistry - Practical Aspects*, ed. B.W.Howard GC CRC Press LLC: USA, 71-102.

Birk,H.-W. and Koepsell,H. (1987). Reaction of Monoclonal Antibodies with Plasma membrane Proteins after Binding on Nitrocellulose: Renaturation of Antigenic Sites and Reduction of Nonspecific Antibody Binding. Anal. Biochem. *164*, 12-22.

Birnboim,H.C. and Doly,J. (1979). A rapid alkaline extraction procedure for screening recombinant plasmid DNA. Nucleic Acids Res. *7*, 1513-1523.

Blackmore,M., Whitwell,C., Morton,J.K., Gray,T.J., and Lock,E.A. (2002). The effects of haloalkene cysteine conjugates on cytosolic free calcium levels in LLC-PK(1) cells--studies utilising digital imaging fluorescence microscopy. Toxicology *175*, 143-152.

Bloom,J. and Pagano,M. (2005). Experimental tests to definitively determine ubiquitylation of a substrate. Methods Enzymol. *399*, 249-266.

Bootman,M.D. and Berridge,M.J. (1996). Subcellular Ca^{2+} signals underlying waves and graded responses in HeLa cells. Curr. Biol. *6*, 855-865.

Botbol,V. and Scornik,O.A. (1983). Peptide intermediates in the degradation of cellular proteins. Bestatin permits their accumulation in mouse liver in vivo. J. Biol. Chem. *258*, 1942-1949.

Bradford,M.M. (1976). A rapid and sensitive method for the quantification of microgram quantities of protein utilizing the principle of protein-dye binding. Anal. Biochem. *72*, 248-254.

Briggs,L.J., Johnstone,R.W., Elliot,R.M., Xiao,C.Y., Dawson,M., Trapani,J.A., and Jans,D.A. (2001). Novel properties of the protein kinase CK2-site-regulated nuclear- localization sequence of the interferon-induced nuclear factor IFI 16. Biochem. J. *353*, 69-77.

Buchner,K. (2000). The role of protein kinase C in the regulation of cell growth and in signalling to the cell nucleus. J. Cancer Res. Clin. Oncol. *126*, 1-11.

Bui,M., Myers,J.E., and Whittaker,G.R. (2002). Nucleo-cytoplasmic localization of influenza virus nucleoprotein depends on cell density and phosphorylation. Virus Res. *84*, 37-44.

Chakraborty,M., Chatterjee,D., Gorelick,F.S., and Baron,R. (1994). Cell cycle-dependent and kinase-specific regulation of the apical Na/H exchanger and the Na,K-ATPase in the kidney cell line LLC-PK1 by calcitonin. Proc. Natl. Acad. Sci. U. S. A *91*, 2115-2119.

Chakraborty,M., Chatterjee,D., Kellokumpu,S., Rasmussen,H., and Baron,R. (1991). Cell cycle-dependent coupling of the calcitonin receptor to different G proteins. Science *251*, 1078-1082.

Chandra,S., Fewtrell,C., Millard,P.J., Sandison,D.R., Webb,W.W., and Morrison,G.H. (1994). Imaging of total intracellular calcium and calcium influx and efflux in individual resting and stimulated tumor mast cells using ion microscopy. J. Biol. Chem. *269*, 15186-15194.

Chen,C., Zhou,Z., Guo,P., and Dong,J.T. (2007). Proteasomal degradation of the KLF5 transcription factor through a ubiquitin-independent pathway. FEBS Lett. *581*, 1124-1130.

Ciechanover,A., Finley,D., and Varshavsky,A. (1984). The ubiquitin-mediated proteolytic pathway and mechanisms of energy-dependent intracellular protein degradation. J. Cell Biochem. *24*, 27-53.

Cingolani,G., Bednenko,J., Gillespie,M.T., and Gerace,L. (2002a). Molecular basis for the recognition of a nonclassical nuclear localization signal by importin beta. Mol. Cell *10*, 1345-1353.

Cingolani,G., Petosa,C., Weis,K., and Muller,C.W. (1999a). Structure of importin-beta bound to the IBB domain of importin-alpha. Nature *399*, 221-229.

Conti,E. and Kuriyan,J. (2000). Crystallographic analysis of the specific yet versatile recognition of distinct nuclear localization signals by karyopherin alpha. Structure. *8*, 329-338.

Conti,E., Uy,M., Leighton,L., Blobel,G., and Kuriyan,J. (1998). Crystallographic analysis of the recognition of a nuclear localization signal by the nuclear import factor karyopherin alpha. Cell *94*, 193-204.

Cook,A., Bono,F., Jinek,M., and Conti,E. (2007). Structural biology of nucleocytoplasmic transport. Annu. Rev. Biochem. *76*, 647-671.

Craig,A.G., Engstrom,A., Bennich,H., Hoffmann-Posorske,E., and Meyer,H.E. (1991). Plasma desorption mass spectrometry of phosphopeptides: an investigation to determine the feasibility of quantifying the degree of phosphorylation. Biol. Mass Spectrom. *20*, 565-574.

Craske,M., Takeo,T., Gerasimenko,O., Vaillant,C., Torok,K., Petersen,O.H., and Tepikin,A.V. (1999). Hormone-induced secretory and nuclear translocation of calmodulin: oscillations of calmodulin concentration with the nucleus as an integrator. Proc. Natl. Acad. Sci. U. S. A *96*, 4426-4431.

Cronshaw,J.M., Krutchinsky,A.N., Zhang,W., Chait,B.T., and Matunis,M.J. (2002). Proteomic analysis of the mammalian nuclear pore complex. J. Cell Biol. *158*, 915-927.

Cuerrier,D., Moldoveanu,T., and Davies,P.L. (2005). Determination of peptide substrate specificity for mu-calpain by a peptide library-based approach: the importance of primed side interactions. J. Biol. Chem. *280*, 40632-40641.

Dang,C.V. and Lee,W.M. (1988). Identification of the human c-myc protein nuclear translocation signal. Mol. Cell Biol. *8*, 4048-4054.

Dawson,W.D. and Cook,J.S. (1987). Parallel changes in amino acid transport and protein kinase C localization in LLC-PK$_1$ cells treated with TPA or diradylglycerols. J. Cell. Physiol. *132*, 104-110.

Dono,R., James,D., and Zeller,R. (1998). A GR-motif functions in nuclear accumulation of the large FGF-2 isoforms and interferes with mitogenic signalling. Oncogene *16*, 2151-2158.

Dower,W.J., Miller,J.F., and Ragsdale,C.W. (1988). High efficiency transformation of E. coli by high voltage electroporation. Nucleic Acids Res. *16*, 6127-6145.

Eldar,H., Ben-Chaim,J., and Livneh,E. (1992). Deletions in the regulatory or kinase domains of protein kinase C-a cause association with the cell nucleus. Exp. cell res. *202*, 259-266.

Fan,X.C. and Steitz,J.A. (1998). HNS, a nuclear-cytoplasmic shuttling sequence in HuR. Proc. Natl. Acad. Sci. U. S. A *95*, 15293-15298.

Feldherr,C.M. and Akin,D. (1993). Regulation of nuclear transport in proliferating and quiescent cells. Exp. Cell Res. *205*, 179-186.

Feldherr,C.M. and Akin,D. (1994). Variations in signal-mediated nuclear transport during the cell cycle in BALB/c 3T3 cells. Exp. Cell Res. *215*, 206-210.

Feldherr,C.M., Akin,D., and Cohen,R.J. (2001). Regulation of functional nuclear pore size in fibroblasts. J. Cell Sci. *114*, 4621-4627.

Fischer,R., Wei,Y., Anagli,J., and Berchtold,M.W. (1996). Calmodulin binds to and inhibits GTP binding of the ras-like GTPase Kir/Gem. J. Biol. Chem. *271*, 25067-25070.

Fischer,U., Huber,J., Boelens,W.C., Mattaj,I.W., and Luhrmann,R. (1995a). The HIV-1 Rev activation domain is a nuclear export signal that accesses an export pathway used by specific cellular RNAs. Cell *82*, 475-483.

Fontes,M.R., Teh,T., and Kobe,B. (2000). Structural basis of recognition of monopartite and bipartite nuclear localization sequences by mammalian importin-alpha. J. Mol. Biol. *297*, 1183-1194.

Fornerod,M., Ohno,M., Yoshida,M., and Mattaj,I.W. (1997). CRM1 is an export receptor for leucine-rich nuclear export signals. Cell *90*, 1051-1060.

Franca-Koh,J., Yeo,M., Fraser,E., Young,N., and Dale,T.C. (2002). The regulation of glycogen synthase kinase-3 nuclear export by Frat/GBP. J. Biol. Chem. *277*, 43844-43848.

Freeman,T.C., Heavens,R.P., Dyer,J., Sirinathsinghji,D.J.S., and Shirazi-Beechey,S.P. (1992). The expression of the Na^+/glucose cotransporter in the lamb small intestine. Biochemical Society Transactions *20*, 1868.

Fried,H. and Kutay,U. (2003). Nucleocytoplasmic transport: taking an inventory. Cell Mol. Life Sci. *60*, 1659-1688.

Friedrich,P. and Bozoky,Z. (2005). Digestive versus regulatory proteases: on calpain action in vivo. Biol. Chem. *386*, 609-612.

Gerasimenko,O.V., Gerasimenko,J.V., Tepikin,A.V., and Petersen,O.H. (1996). Calcium transport pathways in the nucleus. Pflügers Arch. - Eur. J. Physiol. *432*, 1-6.

Gershoni,J.M. and Palade,G.E. (1983). Protein blotting: principles and applications. Anal. Biochem. *131*, 1-15.

Gilchrist,D. and Rexach,M. (2003). Molecular basis for the rapid dissociation of nuclear localization signals from karyopherin alpha in the nucleoplasm. J. Biol. Chem. *278*, 51937-51949.

Goldfarb,D.S., Corbett,A.H., Mason,D.A., Harreman,M.T., and Adam,S.A. (2004). Importin alpha: a multipurpose nuclear-transport receptor. Trends Cell Biol. *14*, 505-514.

Goll,D.E., Thompson,V.F., Li,H., Wei,W., and Cong,J. (2003). The calpain system. Physiol Rev. *83*, 731-801.

Gorlich,D. (1998). Transport into and out of the cell nucleus. EMBO J. *17*, 2721-2727.

Gottardi,C.J., Arpin,M., Fanning,A.S., and Louvard,D. (1996). The junction-associated protein, zonula occludens-1, localizes to the nucleus before the maturation and during the remodeling of cell-cell contacts. Proc. Natl. Acad. Sci U. S. A *93*, 10779-10784.

Graham,F.L., Smiley,J., Russell,W.C., and Nairn,R. (1977). Characteristics of a human cell line transformed by DNA from human adenovirus type 5. J. Gen. Virol. *36*, 59-74.

Grant,S.G., Jessee,J., Bloom,F.R., and Hanahan,D. (1990). Differential plasmid rescue from transgenic mouse DNAs into Escherichia coli methylation-restriction mutants. Proc. Natl. Acad. Sci. U. S. A *87*, 4645-4649.

Gustin,K.E. and Sarnow,P. (2001). Effects of poliovirus infection on nucleo-cytoplasmic trafficking and nuclear pore complex composition. EMBO J. *20*, 240-249.

Han,Y., Weinman,S., Boldogh,I., Walker,R.K., and Brasier,A.R. (1999). Tumor necrosis factor-alpha-inducible IkappaBalpha proteolysis mediated by cytosolic m-calpain. A mechanism parallel to the ubiquitin-proteasome pathway for nuclear factor-kappab activation. J. Biol. Chem. *274*, 787-794.

Henderson,B.R. and Eleftheriou,A. (2000). A comparison of the activity, sequence specificity, and CRM1-dependence of different nuclear export signals. Exp Cell Res *256*, 213-224.

Henkel,T., Zabel,U., van,Z.K., Muller,J.M., Fanning,E., and Baeuerle,P.A. (1992). Intramolecular masking of the nuclear location signal and dimerization domain in the precursor for the p50 NF-kappa B subunit. Cell *68*, 1121-1133.

Hershko,A. (2005). The ubiquitin system for protein degradation and some of its roles in the control of the cell division cycle. Cell Death. Differ. *12*, 1191-1197.

Hershko,A. and Ciechanover,A. (1998). The ubiquitin system. Annu. Rev. Biochem. *67*, 425-479.

Hershko,A., Ciechanover,A., and Varshavsky,A. (2000). Basic Medical Research Award. The ubiquitin system. Nat. Med. *6*, 1073-1081.

Hidaka,H., Sasaki,Y., Tanaka,T., Endo,T., Ohno,S., Fujii,Y., and Nagata,T. (1981). N-(6-aminohexyl)-5-chloro-1-naphthalenesulfonamide, a calmodulin antagonist, inhibits cell proliferation. Proc. Natl. Acad. Sci. U. S. A *78*, 4354-4357.

Hjerpe,R. and Rodriguez,M.S. (2008). Efficient approaches for characterizing ubiquitinated proteins. Biochem. Soc. Trans. *36*, 823-827.

Hodel,A.E., Harreman,M.T., Pulliam,K.F., Harben,M.E., Holmes,J.S., Hodel,M.R., Berland,K.M., and Corbett,A.H. (2006). Nuclear localization signal receptor affinity correlates with in vivo localization in Saccharomyces cerevisiae. J. Biol. Chem. *281*, 23545-23556.

Hull,R.N., Cherry,W.R., and Weaver,G.W. (1976). The origin and characteristics of a pig kidney cell strain, LLC-PK1. IN VITRO *12*, 670-677.

Huxford,T., Malek,S., and Ghosh,G. (1999). Structure and mechanism in NF-kappa B/I kappa B signaling. Cold Spring Harb. Symp. Quant. Biol. *64*, 533-540.

Ibrahim,M., Upreti,R.K., and Kidwai,A.M. (1994). Calpain from rat intestinal epithelial cells: age-dependent dynamics during cell differentiation. Mol. Cell Biochem. *131*, 49-59.

Ikuta,T., Kobayashi,Y., and Kawajiri,K. (2004). Cell density regulates intracellular localization of Aryl hydrocarbon receptor. J. Biol. Chem. *279*, 19209-19216.

Ishidate,T., Yoshihara,S., Kawasaki,Y., Roy,B.C., Toyoshima,K., and Akiyama,T. (1997). Identification of a novel nuclear localization signal in Sam68. FEBS Lett. *409*, 237-241.

Islas,S., Vega,J., Ponce,L., and González-Mariscal,L. (2002). Nuclear localization of the tight junction protein ZO-2 in epithelial cells. Exp Cell Res *274*, 138-148.

Jackman,J. and O'Connor,P.M. (2001). Methods for synchronizing cells at specific stages of the cell cycle. Curr. Protoc. Cell Biol. *Chapter 8*, Unit 8.3.

Jans,D.A., Xiao,C.Y., and Lam,M.H. (2000). Nuclear targeting signal recognition: a key control point in nuclear transport? BioEssays *22*, 532-544.

Jariel-Encontre,I., Pariat,M., Martin,F., Carillo,S., Salvat,C., and Piechaczyk,M. (1995). Ubiquitinylation is not an absolute requirement for degradation of c-Jun protein by the 26 S proteasome. J. Biol. Chem. *270*, 11623-11627.

Jiang,W., Prokopenko,O., Wong,L., Inouye,M., and Mirochnitchenko,O. (2005). IRIP, a new ischemia/reperfusion-inducible protein that participates in the regulation of transporter activity. Mol. Cell Biol. *25*, 6496-6508.

Jin,Y., Lee,H., Zeng,S.X., Dai,M.S., and Lu,H. (2003). MDM2 promotes p21waf1/cip1 proteasomal turnover independently of ubiquitylation. EMBO J. *22*, 6365-6377.

Johnson,C., Van,A.D., and Hope,T.J. (1999). An N-terminal nuclear export signal is required for the nucleocytoplasmic shuttling of IkappaBalpha. EMBO J. *18*, 6682-6693.

Kaffman,A., Rank,N.M., O'Neill,E.M., Huang,L.S., and O'Shea,E.K. (1998). The receptor Msn5 exports the phosphorylated transcription factor Pho4 out of the nucleus. Nature *396*, 482-486.

Kaiser,P. and Tagwerker,C. (2005). Is this protein ubiquitinated? Methods Enzymol. *399*, 243-248.

Kalderon,D., Roberts,B.L., Richardson,W.D., and Smith,A.E. (1984). A short amino acid sequence able to specify nuclear location. Cell *39*, 499-509.

Karlsson,M., Mathers,J., Dickinson,R.J., Mandl,M., and Keyse,S.M. (2004). Both nuclear-cytoplasmic shuttling of the dual specificity phosphatase MKP-3 and its ability to anchor MAP kinase in the cytoplasm are mediated by a conserved nuclear export signal. J. Biol. Chem. *279*, 41882-41891.

Katz,J.P., Perreault,N., Goldstein,B.G., Chao,H.H., Ferraris,R.P., and Kaestner,K.H. (2004). Foxl1 null mice have abnormal intestinal epithelia, postnatal growth retardation, and defective intestinal glucose uptake. Am. J. Physiol Gastrointest. Liver Physiol *287*, G856-G864.

Kekuda,R., Saha,P., and Sundaram,U. (2008). Role of Sp1 and HNF1 transcription factors in SGLT1 regulation during chronic intestinal inflammation. Am. J. Physiol. Gastrointest. Liver Physiol. *294*, G1354-G1361.

Kobayashi,T., Zhang,G., Lee,B.-J., Baba,S., Yamashita,M., Kamitani,W., Yanai,H., Tomonaga,K., and Ikuta,K. (2003). Modulation of Borna disease virus phosphoprotein nuclear localization by the viral protein X encoded in the overlapping open reading frame. J. Virol. *77*, 8099-8107.

Koepsell,H. and Spangenberg,J. (1994). Function and presumed molecular structure of Na(+)-D-glucose cotransport systems. J. Membr. Biol. *138*, 1-11.

Korn,T., Kühlkamp,T., Track,C., Schatz,I., Baumgarten,K., Gorboulev,V., and Koepsell,H. (2001). The plasma membrane-associated protein RS1 decreases transcription of the transporter SGLT1 in confluent LLC-PK$_1$ cells. J. Biol. Chem. *276*, 45330-45340.

Kroiss,M., Leyerer,M., Gorboulev,V., Kühlkamp,T., Kipp,H., and Koepsell,H. (2006). Transporter regulator RS1 *(RSC1A1)* coats the *trans*-Golgi network and migrates into the nucleus. Am. J. Physiol. Renal Physiol. *291*, F1201-F1212.

Kudo,N., Khochbin,S., Nishi,K., Kitano,K., Yanagida,M., Yoshida,M., and Horinouchi,S. (1997). Molecular cloning and cell cycle-dependent expression of mammalian CRM1, a protein involved in nuclear export of proteins. J. Biol. Chem. *272*, 29742-29751.

Kudo,N., Matsumori,N., Taoka,H., Fujiwara,D., Schreiner,E.P., Wolff,B., Yoshida,M., and Horinouchi,S. (1999). Leptomycin B inactivates CRM1/exportin 1 by covalent modification at a cysteine residue in the central conserved region. Proc. Natl. Acad. Sci. USA *96*, 9112-9117.

Kuge,S., Arita,M., Murayama,A., Maeta,K., Izawa,S., Inoue,Y., and Nomoto,A. (2001). Regulation of the yeast Yap1p nuclear export signal is mediated by redox signal-induced reversible disulfide bond formation. Mol. Cell Biol. *21*, 6139-6150.

Kutay,U., Bischoff,F.R., Kostka,S., Kraft,R., and Gorlich,D. (1997a). Export of importin alpha from the nucleus is mediated by a specific nuclear transport factor. Cell *90*, 1061-1071.

Kutay,U. and Güttinger,S. (2005). Leucine-rich nuclear-export signals: born to be weak. Trends Cell Biol. *15*, 121-124.

Kutay,U., Lipowsky,G., Izaurralde,E., Bischoff,F.R., Schwarzmaier,P., Hartmann,E., and Gorlich,D. (1998). Identification of a tRNA-specific nuclear export receptor. Mol. Cell *1*, 359-369.

Kutty,R.K., Chen,S., Samuel,W., Vijayasarathy,C., Duncan,T., Tsai,J.-Y., Fariss,R.N., Carper,D., Jaworski,C., and Wiggert,B. (2006). Cell density-dependent nuclear/cytoplasmic localization of NORPEG (RAI14) protein. Biochem. Biophys. Res. Commun. *345*, 1333-1341.

Laemmli,U.K. (1970). Cleavage of Structural Proteins during the Assembly of the Head of Bacteriophage T4. Nature *227*, 680-685.

Lam,M.H., Briggs,L.J., Hu,W., Martin,T.J., Gillespie,M.T., and Jans,D.A. (1999). Importin beta recognizes parathyroid hormone-related protein with high affinity and mediates its nuclear import in the absence of importin alpha. J. Biol. Chem. *274*, 7391-7398.

Lambotte,S., Veyhl,M., Köhler,M., Morrison-Shetlar,A.I., Kinne,R.K.H., Schmid,M., and Koepsell,H. (1996). The human gene of a protein that modifies Na^+-D-glucose co-transport. DNA Cell Biol. *15*, 769-777.

Lange,A., Mills,R.E., Lange,C.J., Stewart,M., Devine,S.E., and Corbett,A.H. (2007). Classical nuclear localization signals: definition, function, and interaction with importin a. J. Biol. Chem. *282*, 5101-5105.

Leach,K.L., Powers,E.A., Ruff,V.A., Jaken,S., and Kaufmann,S. (1989). Type 3 protein kinase C localization to the nuclear envelope of phorbol ester-treated NIH 3T3 cells. J. Cell Biol. *109*, 685-695.

Lee,D.H. and Goldberg,A.L. (1998). Proteasome inhibitors: valuable new tools for cell biologists. Trends Cell Biol. *8*, 397-403.

Lee,S., Chen,D.Y.T., Humphrey,J.S., Gnarra,J.R., Linehan,W.M., and Klausner,R.D. (1996). Nuclear/cytoplasmic localization of the von Hippel-Lindau tumor suppressor gene product is determined by cell density. Proc. Natl. Acad. Sci. U. S. A. *93*, 1770-1775.

Lee,S.J. *et al.* (2003a). The structure of importin-beta bound to SREBP-2: nuclear import of a transcription factor. Science *302*, 1571-1575.

Lee,S.J., Sekimoto,T., Yamashita,E., Nagoshi,E., Nakagawa,A., Tanaka,H., Yoneda,Y., and Tsukihara,T. (2003c). Crystallization and preliminary crystallographic analysis of the importin-beta-SREBP-2 complex. Acta Crystallogr. D. Biol. Crystallogr. *59*, 1866-1868.

Lee,Y.J., Heo,J.S., Suh,H.N., Lee,M.Y., and Han,H.J. (2007). Interleukin-6 stimulates alpha-MG uptake in renal proximal tubule cells: involvement of STAT3, PI3K/Akt, MAPKs, and NF-kappaB. Am. J. Physiol Renal Physiol *293*, F1036-F1046.

Lever,J.E. (1986). Expression of differentiated functions in kidney epithelial cell lines. Miner. Electrolyte Metab *12*, 14-19.

Leyerer,M. Identification and characterization of nuclear localization signal of pRS1 protein. 2007. Anatomisches Institut der Universität Würzburg (Lehrstuhl I). Ref Type: Thesis/Dissertation

Li,X., Lonard,D.M., Jung,S.Y., Malovannaya,A., Feng,Q., Qin,J., Tsai,S.Y., Tsai,M.J., and O'Malley,B.W. (2006). The SRC-3/AIB1 coactivator is degraded in a ubiquitin- and ATP-independent manner by the REGgamma proteasome. Cell *124*, 381-392.

Lin,Y.T. and Yen,P.H. (2006). A novel nucleocytoplasmic shuttling sequence of DAZAP1, a testis-abundant RNA-binding protein. RNA. *12*, 1486-1493.

Lischka,P., Sorg,G., Kann,M., Winkler,M., and Stamminger,T. (2003). A nonconventional nuclear localization signal within the UL84 protein of human cytomegalovirus mediates nuclear import via the importin alpha/beta pathway. J. Virol. *77*, 3734-3748.

Lixin,R., Efthymiadis,A., Henderson,B., and Jans,D.A. (2001). Novel properties of the nucleolar targeting signal of human angiogenin. Biochem. Biophys. Res. Commun. *284*, 185-193.

Lusk,C.P., Makhnevych,T., and Wozniak,R.W. (2004). New ways to skin a kap: mechanisms for controlling nuclear transport. Biochem. Cell Biol. *82*, 618-625.

Macara,I.G. (2001a). Transport into and out of the nucleus. Microbiol. Mol. Biol. Rev. *65*, 570-94.

Martelli,A.M., Evangelisti,C., Nyakern,M., and Manzoli,F.A. (2006). Nuclear protein kinase C. Biochim. Biophys. Acta *1761*, 542-551.

Martelli,A.M., Sang,N., Borgatti,P., Capitani,S., and Neri,L.M. (1999). Multiple biological responses activated by nuclear protein kinase C. J. Cell Biochem. *74*, 499-521.

Martin,M.G., Wang,J., Solorzano-Vargas,R.S., Lam,J.T., Turk,E., and Wright,E.M. (2000). Regulation of the human Na^+-glucose cotransporter gene, *SGLT1*, by HNF-1 and Sp1. Am. J. Physiol. Gastrointest. Liver Physiol. *278*, G591-G603.

Matsuura,Y., Lange,A., Harreman,M.T., Corbett,A.H., and Stewart,M. (2003a). Structural basis for Nup2p function in cargo release and karyopherin recycling in nuclear import. EMBO J. *22*, 5358-5369.

Matsuura,Y. and Stewart,M. (2005). Nup50/Npap60 function in nuclear protein import complex disassembly and importin recycling. EMBO J. *24*, 3681-3689.

Matunis,M.J., Wu,J., and Blobel,G. (1998). SUMO-1 modification and its role in targeting the Ran GTPase-activating protein, RanGAP1, to the nuclear pore complex. J Cell Biol. *140*, 499-509.

McBride,K.M., Banninger,G., McDonald,C., and Reich,N.C. (2002). Regulated nuclear import of the STAT1 transcription factor by direct binding of importin-alpha. EMBO J. *21*, 1754-1763.

Melen,K., Kinnunen,L., and Julkunen,I. (2001). Arginine/lysine-rich structural element is involved in interferon-induced nuclear import of STATs. J. Biol. Chem. *276*, 16447-16455.

Merril,C.R., Dunau,M.L., and Goldman,D. (1981). A rapid sensitive silver stain for polypeptides in polyacrylamide gels. Anal. Biochem. *110*, 201-207.

Merril,C.R. and Pratt,M.E. (1986). A silver stain for the rapid quantitative detection of proteins or nucleic acids on membranes or thin layer plates. Anal. Biochem. *156*, 96-110.

Michael,W.M. (2000). Nucleocytoplasmic shuttling signals: two for the price of one. Trends Cell Biol. *10*, 46-50.

Michael,W.M., Choi,M., and Dreyfuss,G. (1995). A nuclear export signal in hnRNP A1: a signal-mediated, temperature-dependent nuclear protein export pathway. Cell *83*, 415-422.

Mullin,J.M., Weibel,J., Diamond,L., and Kleinzeller,A. (1980). Sugar transport in the LLC-PK1 renal epithelial cell line: similarity to mammalian kidney and the influence of cell density. J. Cell. Physiol. *104*, 375-389.

Murakami,Y., Matsufuji,S., Kameji,T., Hayashi,S., Igarashi,K., Tamura,T., Tanaka,K., and Ichihara,A. (1992). Ornithine decarboxylase is degraded by the 26S proteasome without ubiquitination. Nature *360*, 597-599.

Nadler,S.G., Tritschler,D., Haffar,O.K., Blake,J., Bruce,A.G., and Cleaveland,J.S. (1997). Differential expression and sequence-specific interaction of karyopherin alpha with nuclear localization sequences. J. Biol. Chem. *272*, 4310-4315.

Nagoshi,E., Imamoto,N., Sato,R., and Yoneda,Y. (1999). Nuclear import of sterol regulatory element-binding protein-2, a basic helix-loop-helix-leucine zipper (bHLH-Zip)-containing transcription factor, occurs through the direct interaction of importin beta with HLH-Zip. Mol. Biol. Cell *10*, 2221-2233.

Nagoshi,E. and Yoneda,Y. (2001). Dimerization of sterol regulatory element-binding protein 2 via the helix-loop-helix-leucine zipper domain is a prerequisite for its nuclear localization mediated by importin beta. Mol. Cell Biol. *21*, 2779-2789.

Nakielny,S. and Dreyfuss,G. (1999). Transport of proteins and RNAs in and out of the nucleus. Cell *99*, 677-690.

Nigam,S.K., Rodriguez-Boulan,E., and Silver,R.B. (1992). Changes in intracellular calcium during the development of epithelial polarity and junctions. Proc. Natl. Acad. Sci. U. S. A. *89*, 6162-6166.

Nishi,K., Yoshida,M., Fujiwara,D., Nishikawa,M., Horinouchi,S., and Beppu,T. (1994). Leptomycin B targets a regulatory cascade of crm1, a fission yeast nuclear protein, involved in control of higher order chromosome structure and gene expression. J. Biol. Chem. *269*, 6320-6324.

Oakley,B.R., Kirsch,D.R., and Morris,N.R. (1980). A simplified ultrasensitive silver stain for detecting proteins in polyacrylamide gels. Anal. Biochem. *105*, 361-363.

Oesterreicher,T.J., Leeper,L.L., Finegold,M.J., Darlington,G.J., and Henning,S.J. (1998). Intestinal maturation in mice lacking CCAAT/enhancer-binding protein alpha (C/EPBalpha). Biochem. J. *330 (Pt 3)*, 1165-1171.

Ohno,M., Fornerod,M., and Mattaj,I.W. (1998). Nucleocytoplasmic transport: the last 200 nanometers. Cell *92*, 327-336.

Ossareh-Nazari,B., Bachelerie,F., and Dargemont,C. (1997). Evidence for a role of CRM1 in signal-mediated nuclear protein export. Science *278*, 141-144.

Osswald,C., Baumgarten,K., Stümpel,F., Gorboulev,V., Akimjanova,M., Knobeloch,K.-P., Horak,I., Kluge,R., Joost,H.-G., and Koepsell,H. (2005). Mice without the regulator gene *Rsc1A1* exhibit increased Na^+-D-glucose cotransport in small intestine and develop obesity. Mol. Cell Biol. *25*, 78-87.

Parys,J.B., De,S.H., and Borghgraef,R. (1986). Calcium transport systems in the LLC-PK1 renal epithelial established cell line. Biochim. Biophys. Acta *888*, 70-81.

Pemberton,L.F. and Paschal,B.M. (2005). Mechanisms of receptor-mediated nuclear import and nuclear export. Traffic *6*, 187-198.

Peng,H. and Lever,J.E. (1995). Post-transcriptional regulation of Na^+/glucose cotransporter (*SGLT1*) gene expression in LLC-PK$_1$ cells - increased message stability after cyclic AMP elevation or differentiation inducer treatment. J. Biol. Chem. *270*, 20536-20542.

Petersen,O.H., Burdakov,D., and Tepikin,A.V. (1999). Polarity in intracellular calcium signaling. BioEssays *21*, 851-860.

Petridou,S., Maltseva,O., Spanakis,S., and Masur,S.K. (2000). TGF-b receptor expression and smad2 localization are cell density dependent in fibroblasts. Invest. Ophthalmol. Vis. Sci. *41*, 89-95.

Pickart,C.M. and Cohen,R.E. (2004). Proteasomes and their kin: proteases in the machine age. Nat. Rev. Mol. Cell Biol. *5*, 177-187.

Pollard,V.W., Michael,W.M., Nakielny,S., Siomi,M.C., Wang,F., and Dreyfuss,G. (1996). A novel receptor-mediated nuclear protein import pathway. Cell *86*, 985-994.

Poon,I.K.H. and Jans,D.A. (2005). Regulation of nuclear transport: central role in development and transformation? Traffic *6*, 173-186.

Poppe,R., Karbach,U., Gambaryan,S., Wiesinger,H., Lutzenburg,M., Kraemer,M., Witte,O.W., and Koepsell,H. (1997). Expression of the Na^+-D-glucose cotransporter SGLT1 in neurons. J. Neurochem. *69*, 84-94.

Potter,D.A., Srirangam,A., Fiacco,K.A., Brocks,D., Hawes,J., Herndon,C., Maki,M., Acheson,D., and Herman,I.M. (2003). Calpain regulates enterocyte brush border actin assembly and pathogenic Escherichia coli-mediated effacement. J. Biol. Chem. *278*, 30403-30412.

Prieve,M.G., Guttridge,K.L., Munguia,J.E., and Waterman,M.L. (1996). The nuclear localization signal of lymphoid enhancer factor-1 is recognized by two differentially expressed Srp1-nuclear localization sequence receptor proteins. J. Biol. Chem. *271*, 7654-7658.

Pyhtila,B. and Rexach,M. (2003). A gradient of affinity for the karyopherin Kap95p along the yeast nuclear pore complex. J. Biol. Chem. *278*, 42699-42709.

Rabilloud,T. (1990). Mechanisms of protein silver staining in polyacrylamide gels: a 10-year synthesis. Electrophoresis *11*, 785-794.

Rechsteiner,M. and Rogers,S.W. (1996). PEST sequences and regulation by proteolysis. Trends Biochem. Sci. *21*, 267-271.

Reinhardt,J., Veyhl,M., Wagner,K., Gambaryan,S., Dekel,C., Akhoundova,A., Korn,T., and Koepsell,H. (1999). Cloning and characterization of the transport modifier RS1 from rabbit which was previously assumed to be specific for Na^+-D-glucose cotransport. Biochim. Biophys. Acta *1417*, 131-143.

Rhoads,A.R. and Friedberg,F. (1997). Sequence motifs for calmodulin recognition. FASEB J. *11*, 331-340.

Rhoads,D.B., Rosenbaum,D.H., Unsal,H., Isselbacher,K.J., and Levitsky,L.L. (1998). Circadian periodicity of intestinal Na^+/glucose cotransporter 1 mRNA levels is transcriptionally regulated. J. Biol. Chem. *273*, 9510-9516.

Ribbeck,K. and Gorlich,D. (2001). Kinetic analysis of translocation through nuclear pore complexes. EMBO J. *20*, 1320-1330.

Riviere,Y., Blank,V., Kourilsky,P., and Israel,A. (1991). Processing of the precursor of NF-kappa B by the HIV-1 protease during acute infection. Nature *350*, 625-626.

Robbins,J., Dilworth,S.M., Laskey,R.A., and Dingwall,C. (1991a). Two interdependent basic domains in nucleoplasmin nuclear targeting sequence: identification of a class of bipartite nuclear targeting sequence. Cell *64*, 615-623.

Rogers,S., Wells,R., and Rechsteiner,M. (1986). Amino acid sequences common to rapidly degraded proteins: the PEST hypothesis. Science *234*, 364-368.

Rosenblum,J.S., Pemberton,L.F., Bonifaci,N., and Blobel,G. (1998). Nuclear import and the evolution of a multifunctional RNA-binding protein. J. Cell Biol. *143*, 887-899.

Rosorius,O., Heger,P., Stelz,G., Hirschmann,N., Hauber,J., and Stauber,R.H. (1999). Direct observation of nucleocytoplasmic transport by microinjection of GFP-tagged proteins in living cells. BioTechniques *27*, 350-355.

Roth,A.F., Sullivan,D.M., and Davis,N.G. (1998). A large PEST-like sequence directs the ubiquitination, endocytosis, and vacuolar degradation of the yeast a-factor receptor. J. Cell Biol. *142*, 949-961.

Sachdev,S. and Hannink,M. (1998). Loss of IkBa-mediated control over nuclear import and DNA binding enables oncogenic activation of c-Rel. Mol. Cell. Biol. *18*, 5445-5456.

Sandoval,A., Oviedo,N., Tadmouri,A., Avila,T., De,W.M., and Felix,R. (2006). Two PEST-like motifs regulate Ca2+/calpain-mediated cleavage of the CaVbeta3 subunit and provide important determinants for neuronal Ca2+ channel activity. Eur. J. Neurosci. *23*, 2311-2320.

Saporita,A.J., Zhang,Q., Navai,N., Dincer,Z., Hahn,J., Cai,X., and Wang,Z. (2003). Identification and characterization of a ligand-regulated nuclear export signal in androgen receptor. J. Biol. Chem. *278*, 41998-42005.

Sasaki,T., Kikuchi,T., Yumoto,N., Yoshimura,N., and Murachi,T. (1984). Comparative specificity and kinetic studies on porcine calpain I and calpain II with naturally occurring peptides and synthetic fluorogenic substrates. J. Biol. Chem. *259*, 12489-12494.

Sdek,P., Ying,H., Chang,D.L., Qiu,W., Zheng,H., Touitou,R., Allday,M.J., and Xiao,Z.X. (2005). MDM2 promotes proteasome-dependent ubiquitin-independent degradation of retinoblastoma protein. Mol. Cell *20*, 699-708.

Sherman,M.P., De Noronha,C.M., Heusch,M.I., Greene,S., and Greene,W.C. (2001). Nucleocytoplasmic shuttling by human immunodeficiency virus type 1 Vpr. J. Virol. *75*, 1522-1532.

SHIODA,T., Ohta,T., Isselbacher,K.J., and Rhoads,D.B. (1994). Differentiation-dependent expression of the Na+/glucose cotransporter (SGLT1) in LLC-PK1 cells: Role of protein kinase C activation and ongoing transcription. Proc. Natl. Acad. Sci. USA *91*, 11919-11923.

Shumway,S.D., Maki,M., and Miyamoto,S. (1999). The PEST domain of IkappaBalpha is necessary and sufficient for in vitro degradation by mu-calpain. J. Biol. Chem. *274*, 30874-30881.

Siomi,H. and Dreyfuss,G. (1995). A nuclear localization domain in the hnRNP A1 protein. J. Cell Biol. *129*, 551-560.

Siomi,M.C., Eder,P.S., Kataoka,N., Wan,L., Liu,Q., and Dreyfuss,G. (1997). Transportin-mediated nuclear import of heterogeneous nuclear RNP proteins. J. Cell Biol. *138*, 1181-1192.

Slater,S.J., Kelly,M.B., Taddeo,F.J., Rubin,E., and Stubbs,C.D. (1994). Evidence for discrete diacylglycerol and phorbol ester activator sites on protein kinase C. Differences in effects of 1-alkanol inhibition, activation by phosphatidylethanolamine and calcium chelation. J. Biol. Chem. *269*, 17160-17165.

Sorg,G. and Stamminger,T. (1999). Mapping of Nuclear Localization Signals by Simultaneous Fusion to Green Fluorescent Protein and to ß-Galactosidase. Bio Techniques *26*, 858-862.

Stade,K., Ford,C.S., Guthrie,C., and Weis,K. (1997). Exportin 1 (Crm1p) is an essential nuclear export factor. Cell *90*, 1041-1050.

Stewart,M. (2000). Insights into the molecular mechanism of nuclear trafficking using nuclear transport factor 2 (NTF2). Cell Struct. Funct. *25*, 217-225.

Stewart,M. (2007). Molecular mechanism of the nuclear protein import cycle. Nat. Rev. Mol. Cell Biol. *8*, 195-208.

Stommel,J.M., Marchenko,N.D., Jimenez,G.S., Moll,U.M., Hope,T.J., and Wahl,G.M. (1999) A leucine-rich nuclear export signal in the p53 tetramerization domain: regulation of subcellular localization and p53 activity by NES masking. EMBO J. *18*, 1660-1672.

Sussman,J., Stokoe,D., Ossina,N., and Shtivelman,E. (2001). Protein kinase B phosphorylates AHNAK and regulates its subcellular localization. J. Cell Biol. *154*, 1019-1030.

Suzuki,K., Hata,S., Kawabata,Y., and Sorimachi,H. (2004). Structure, activation, and biology of calpain. Diabetes *53 Suppl 1*, S12-S18.

Switzer,R.C., Merril,C.R., and Shifrin,S. (1979). A highly sensitive silver stain for detecting proteins and peptides in polyacrylamide gels. Anal. Biochem. *98*, 231-237.

Tabatabai,N.M., Blumenthal,S.S., and Petering,D.H. (2005). Adverse effect of cadmium on binding of transcription factor Sp1 to the GC-rich regions of the mouse sodium-glucose cotransporter 1, SGLT1, promoter. Toxicology *207*, 369-382.

Takei,Y., Yamamoto,K., and Tsujimoto,G. (1999). Identification of the sequence responsible for the nuclear localization of human Cdc6. FEBS Lett. *447*, 292-296.

Thomas,T.P., Talwar,H.S., and Anderson,W.B. (1988). Phorbol ester-mediated association of protein kinase C to the nuclear fraction in NIH 3T3 cells. Cancer Res. *48*, 1910-1919.

Thorsness,P.E. and Koshland,D.E., Jr. (1987). Inactivation of isocitrate dehydrogenase by phosphorylation is mediated by the negative charge of the phosphate. J. Biol. Chem. *262*, 10422-10425.

Tiganis,T., Flint,A.J., Adam,S.A., and Tonks,N.K. (1997). Association of the T-cell protein tyrosine phosphatase with nuclear import factor p97. J. Biol. Chem. *272*, 21548-21557.

Tompa,P., Buzder-Lantos,P., Tantos,A., Farkas,A., Szilagyi,A., Banoczi,Z., Hudecz,F., and Friedrich,P. (2004). On the sequential determinants of calpain cleavage. J. Biol. Chem. *279*, 20775-20785.

Tremper-Wells,B. and Vallano,M.L. (2005). Nuclear calpain regulates Ca2+-dependent signaling via proteolysis of nuclear Ca2+/calmodulin-dependent protein kinase type IV in cultured neurons. J. Biol. Chem. *280*, 2165-2175.

Truant,R. and Cullen,B.R. (1999). The arginine-rich domains present in human immunodeficiency virus type 1 Tat and Rev function as direct importin beta-dependent nuclear localization signals. Mol. Cell Biol. *19*, 1210-1217.

Tsuji,L., Takumi,T., Imamoto,N., and Yoneda,Y. (1997). Identification of novel homologues of mouse importin alpha, the alpha subunit of the nuclear pore-targeting complex, and their tissue-specific expression. FEBS Lett. *416*, 30-34.

Tsujinaka,T., Kajiwara,Y., Kambayashi,J., Sakon,M., Higuchi,N., Tanaka,T., and Mori,T. (1988). Synthesis of a new cell penetrating calpain inhibitor (calpeptin). Biochem. Biophys. Res. Commun. *153*, 1201-1208.

Tyagi,R.K., Amazit,L., Lescop,P., Milgrom,E., and Guiochon-Mantel,A. (1998). Mechanisms of progesterone receptor export from nuclei: role of nuclear localization signal, nuclear export signal, and ran guanosine triphosphate. Mol. Endocrinol. *12*, 1684-1695.

Ullman,K.S., Powers,M.A., and Forbes,D.J. (1997). Nuclear export receptors: from importin to exportin. Cell *90*, 967-970.

Valacco,M.P., Varone,C., Malicet,C., Cánepa,E., Iovanna,J.L., and Moreno,S. (2006). Cell growth-dependent subcellular localization of p8. J. Cell. Biochem. *97*, 1066-1079.

Valentin,M., Kühlkamp,T., Wagner,K., Krohne,G., Arndt,P., Baumgarten,K., Weber,W.-M., Segal,A., Veyhl,M., and Koepsell,H. (2000). The transport modifier RS1 is localized at the inner side of the plasma membrane and changes membrane capacitance. Biochim. Biophys. Acta *1468*, 367-380.

Vayro,S., Wood,I.S., Dyer,J., and Shirazi-Beechey,S.P. (2001). Transcriptional regulation of the ovine intestinal Na$^+$/glucose cotransporter *SGLT1* gene. Role of HNF-1 in glucose activation of promoter function. Eur. J. Biochem. *268*, 5460-5470.

Vernaleken,A. *et al.* (2007). Tripeptides of RS1 *(RSC1A1)* inhibit a monosaccharide-dependent exocytotic pathway of Na$^+$-D-glucose cotransporter SGLT1 with high affinity. J. Biol. Chem. *282*, 28501-28513.

Veyhl,M., Keller,T., Gorboulev,V., Vernaleken,A., and Koepsell,H. (2006). RS1*(RSC1A1)* regulates the exocytotic pathway of Na$^+$-D-glucose cotransporter SGLT1. Am. J. Physiol. Renal Physiol. *291*, F1213-F1223.

Veyhl,M., Spangenberg,J., Püschel,B., Poppe,R., Dekel,C., Fritzsch,G., Haase,W., and Koepsell,H. (1993). Cloning of a membrane-associated protein which modifies activity and properties of the Na$^+$-D-glucose cotransporter. J. Biol. Chem. *268*, 25041-25053.

Veyhl,M., Wagner,C.A., Gorboulev,V., Schmitt,B.M., Lang,F., and Koepsell,H. (2003). Downregulation of the Na$^+$-D-glucose cotransporter SGLT1 by protein RS1 (RSC1A1) is dependent on dynamin and protein kinase C. J. Membr. Biol. *196*, 71-81.

Wallace,D.M. (1987). Large- and small-scale phenol extractions. Methods Enzymol. *152*, 33-41.

Wang,K.K., Villalobo,A., and Roufogalis,B.D. (1989). Calmodulin-binding proteins as calpain substrates. Biochem. J. *262*, 693-706.

Wang,P., Palese,P., and O'Neill,R.E. (1997). The NPI-1/NPI-3 (karyopherin a) binding site on the influenza a virus nucleoprotein NP is a nonconventional nuclear localization signal. J. Virol. *71*, 1850-1856.

Weis,K. (2003). Regulating access to the genome: nucleocytoplasmic transport throughout the cell cycle. Cell *112*, 441-451.

Wen,W., Meinkoth,J.L., Tsien,R.Y., and Taylor,S.S. (1995). Identification of a signal for rapid export of proteins from the nucleus. Cell *82*, 463-473.

Wilkinson JM. (1986). Fragmentation of Polypeptides by Enzymatic Methods. In: *Practical Protein Chemistry - A Handbook*, ed. Darbre A John Wiley and Sons Ltd: Great Britain, 121-148.

Willems,A.R., Lanker,S., Patton,E.E., Craig,K.L., Nason,T.F., Mathias,N., Kobayashi,R., Wittenberg,C., and Tyers,M. (1996). Cdc53 targets phosphorylated G1 cyclins for degradation by the ubiquitin proteolytic pathway. Cell *86*, 453-463.

Wood,I.S., Allison,G.G., and Shirazi-Beechey,S.P. Isolation and characterization of a genomic region upstream from the ovine Na$^+$/D-glucose cotransporter (SGLT1) cDNA. Biochem.Biophys.Res.Commun. 257, 533-537. 1999.
Ref Type: Journal (Full)

Yagita,K., Tamanini,F., Yasuda,M., Hoeijmakers,J.H., van der Horst,G.T., and Okamura,H. (2002). Nucleocytoplasmic shuttling and mCRY-dependent inhibition of ubiquitylation of the mPER2 clock protein. EMBO J. *21*, 1301-1314.

Yap,K.L., Kim,J., Truong,K., Sherman,M., Yuan,T., and Ikura,M. (2000). Calmodulin target database. J. Struct. Funct. Genomics *1*, 8-14.

Yet,S.F., Kong,C.T., Peng,H., and Lever,J.E. (1994). Regulation of Na+/glucose cotransporter (SGLT1) mRNA in LLC-PK1 cells. J. Cell. Physiol. *158*, 506-512.

Yoshida,K. and Blobel,G. (2001). The karyopherin Kap142p/Msn5p mediates nuclear import and nuclear export of different cargo proteins. J. Cell Biol. *152*, 729-740.

Zatz,M. and Starling,A. (2005). Calpains and disease. N. Engl. J. Med. *352*, 2413-2423.

Zhang,F., White,R.L., and Neufeld,K.L. (2001). Cell density and phosphorylation control the subcellular localization of adenomatous polyposis coli protein. Mol. Cell Biol. *21*, 8143-8156.

Zhao,Y., Hawes,J., Popov,K.M., Jaskiewicz,J., Shimomura,Y., Crabb,D.W., and Harris,R.A. (1994). Site-directed mutagenesis of phosphorylation sites of the branched chain alpha-ketoacid dehydrogenase complex. J. Biol. Chem. *269*, 18583-18587.

Zhou,F., Hu,J., Ma,H., Harrison,M.L., and Geahlen,R.L. (2006). Nucleocytoplasmic trafficking of the Syk protein tyrosine kinase. Mol. Cell. Biol. *26*, 3478-3491.

Acknowledgements

The work presented here was carried out at the Institute of Anatomy and Cell Biology in the Laboratory of Prof. Dr. Hermann Koepsell at University of Würzburg. My profound gratitude goes to Prof. Dr. Hermann Koepsell who thoroughly supervised my work and helped me to find solutions to the numerous problems that emerged in the course of my work. His confidence in my work demonstrated by the allowance to independently and freely conduct my experiments is greatly acknowledged. I am also indebted to Prof. Dr. Hermann Koepsell for editing my thesis.

I would like to thank Prof. Dr. Roland Benz and Dr. Heike Hermanns for co-supervision of my PhD work on behalf of the Graduate School of Life Sciences. I am indebted to Graduate School of Life Sciences. From all the activities organised by graduate committee, I learned a lot in the course of my study.

I am thankful to Dr. Dmitry Gorbunov, Dr. Valentin Gorboulev and Dr. Alexandra Vernaleken for critical reading of the manuscript and important suggestions.

Through the years of the PhD work a number of people have been kindly helping and supporting me, and I would like to take this opportunity and express my sincere appreciation and gratitude:

To Dr. Alexandra Vernaleken, Dr. Valentin Gorboulev and Dr. Thorsten Keller, for the helpful discussions and introducing me into the used techniques;

To Dr. Marina Leyerer and Dr. Valentin Gorboulev, for generation of the majority of the constructs used in this work;

To Chakravarthi Chintalapati, for performing *in vitro* calmodulin binding assays;

To Irina Schatz, Brigitte Dürner and Ursula Roth, for outstanding and diligent technical assistance;

To Dr. Yvonne Reinders, for performing mass spectrometry analysis;

To Dr. Susanne Kneitz, for performing microarray analysis;

To Dr. med. Christoph Klenk, for introducing me in the field of ubiquitination investigation and kindly providing me with ubiquitin encoding plasmids.

I would like to thank my former and current colleagues at the Institute of Anatomy and Cell Biology who made my time in the laboratory highly enjoyable and joyful. It has been a pleasure to have Chakri, Brigitte, Alexandra and Dmitry as PhD students in the lab. I would like to thank Thorsten, Valentin, Helmut, Maike and Christopher for being so friendly and helpful during my stay.

I am thankful to all my friends who made my stay in Germany highly enjoyable and joyful.

Most deeply, I am grateful to my parents and all my family members for providing me support and love.

I want morebooks!

Buy your books fast and straightforward online - at one of the world's fastest growing online book stores! Environmentally sound due to Print-on-Demand technologies.

Buy your books online at
www.get-morebooks.com

Kaufen Sie Ihre Bücher schnell und unkompliziert online – auf einer der am schnellsten wachsenden Buchhandelsplattformen weltweit! Dank Print-On-Demand umwelt- und ressourcenschonend produziert.

Bücher schneller online kaufen
www.morebooks.de

OmniScriptum Marketing DEU GmbH
Heinrich-Böcking-Str. 6-8
D - 66121 Saarbrücken
Telefax: +49 681 93 81 567-9

info@omniscriptum.com
www.omniscriptum.com

Printed by Books on Demand GmbH, Norderstedt / Germany